Reviews of

71 Physiology, Biochemistry and Pharmacology

formerly
Ergebnisse der Physiologie, biologischen
Chemie und experimentellen Pharmakologie

Editors
R. H. Adrian, Cambridge · E. Helmreich, Würzburg
H. Holzer, Freiburg · R. Jung, Freiburg
K. Kramer, München · O. Krayer, Boston
F. Lynen, München · P. A. Miescher, Genève
H. Rasmussen, Philadelphia · A. E. Renold, Genève
U. Trendelenburg, Würzburg · K. Ullrich, Frankfurt/M.
W. Vogt, Göttingen · A. Weber, Philadelphia

With 70 Figures

Springer-Verlag Berlin Heidelberg GmbH 1974

ISBN 978-3-662-31247-6   ISBN 978-3-540-37864-8 (eBook)
DOI 10.1007/978-3-540-37864-8

Contents

List of Contributors

DECKER, K., Prof. Dr., Biochemisches Institut der Universität 7800 Freiburg/Federal Republic of Germany

ISSELBACHER, K. J., Prof. Dr., Harvard Medical School, The Massachusetts General Hospital, Boston, MA 02114/USA

KEPPLER, D., Prof. Dr., Biochemisches Institut der Universität 7800 Freiburg/Federal Republic of Germany

OCKNER, R. K., Dr., Department of Medicine, University of California, School of Medicine, San Francisco, CA 94143/USA

SIMON, E., Prof. Dr., Max-Planck-Institut für Physiologische und Klinische Forschung, W. G. Kerckhoff Institut, 6350 Bad Nauheim/Federal Republic of Germany

Rev. Physiol.Biochem. Pharmacol., Vol. 71
© by Springer Verlag 1974

Temperature Regulation: The Spinal Cord as a Site of Extrahypothalamic Thermoregulatory Functions

ECKHART SIMON*

Contents

* Max-Planck-Institut für Physiologische und Klinische Forschung— W. G. Kerckhoff-Institut, D-6350 Bad Nauheim, Germany.

The investigations concerning the role of central nervous structures in temperature regulation may be considered as having proceeded during the last decade along two converging conceptual lines which have originally emerged from opposing points of view. One of these concepts which can be traced to the classical discoveries of Barbour (1912) and of Isenschmid and Krehl (1912) had proposed that the hypothalamus represents the only center of temperature regulation existing within the central nervous system. Both the control of the thermoregulatory effector activities and the perception of core temperature as the controlled variable were exclusively ascribed to this particular section of the brain stem. The second concept had been introduced in the discussion about the central nervous functions in temperature regulation by Thauer (1939). It emphasized the functional character of nervous control centers and proposed that the specific thermoregulatory control functions, though governed by the hypothalamus, are established in principle by structures widespread in the central nervous system.

The numerous experimental efforts which have been stimulated by these opposing concepts during the last ten years have led to their gradual approach and eventual coalescence. The analyses of hypothalamic functions in temperature regulation have positively confirmed the great importance of this section of the brain stem as a control center and a central temperature sensor, but have also indicated that the hypothalamus alone cannot account for the entire information about deep body temperature. On the other hand, substantial evidence has accumulated that extrahypothalamic central nervous structures contribute significantly to the regulation of body temperature. The experiments which have demonstrated the existence of thermosensitive structures in the spinal cord have been comprehensively described by Thauer (1970). The crucial role of this discovery in definitely establishing central extrahypothalamic thermoregulatory functions has been pointed out recently by Thauer and Simon (1972). It is the aim of this report to review the investigations on the spinal cord and those related studies which have formed the experimental basis for the concept of multiple representation of thermoregulatory control functions in the central nervous system.

1. Spinal Mechanisms of Temperature Regulation

1.1. Extrahypothalamic Deep Body Thermosensitivity in the Spinal Canal

The investigations which have led to the discovery of thermosensitive structures within the spinal cord may be reckoned among the numerous experimental efforts to establish the function of extrahypothalamic deep body thermoreceptors in temperature regulation. These efforts which were characterized by Thauer (1962) as the search for the "glomus caroticum of temperature regulation" had been stimulated by conflicting results concerned with the thermoreceptive function of the hypothalamus.

Among the investigations which had established the predominant role of the hypothalamus in temperature regulation, most examined the effects of lesions in or transections of the brain stem (for References see Hardy, 1961). There were only a few fundamental reports which described thermoregulatory reactions induced by thermal stimulation of the brain stem (BARBOUR, 1912; HASHIMOTO, 1915; PRINCE and HAHN, 1918) or of the hypothalamus (MAGOUN et al., 1938; HEMINGWAY et al., 1940; BEATON et al., 1941). Proceeding from these experiments it was taken for granted that core temperature was perceived exclusively by temperature sensors located in the preoptic and anterior hypothalamic region (BENZINGER et al., 1961; HARDY, 1961). However, the systematic exploration of the thermoreceptive function of the hypothalamus which was started by the investigations of STRÖM (1950a, b) had only in part supported this view. On the one hand, the great influence of hypothalamic temperature on the activity of many thermoregulatory effectors was convincingly demonstrated. On the other hand, several observations were made which reinforced some of the early doubts of THAUER (1939) about the exclusive role of the hypothalamus in central temperature perception. For instance, a high degree of hypothalamic heat and cold sensitivity could be demonstrated in conscious dogs by HAMMEL et al. (1960) and by FUSCO et al. (1961). But the gradual lessening of effector activity during sustained hypothalamic temperature displacements seemed to indicate the existence of deep thermoreceptors outside of the thermally stimulated area. In particular, the diversity of findings concerning heat and cold sensitivity increased the doubts about the concept that central temperature reception was restricted to the hypothalamus. While heating of the preoptic and anterior hypothalamic region consistently elicited heat defence reactions (STRÖM, 1950a; VON EULER and SÖDERBERG, 1958; FREEMAN and DAVIS, 1959; HAMMEL et al., 1960; FUSCO et al., 1961), equivocal results were obtained when the effects of cooling were studied. Negative results and weak or transitory effects as well as definite cold defence reactions in response to cooling of the brain or hypothalamus were reported by different investigators (STRÖM, 1950b; FREEMAN and DAVIS, 1959; DONHOFFER et al., 1959; BETZ et al., 1960; BRENDEL, 1960; HAMMEL et al., 1960; LIM, 1960; ANDERSSON et al., 1962; DOWNEY and MOTTRAM, 1962). While these conflicting results still did not question the predominant role of the hypothalamus in central temperature perception, they favoured theories of temperature regulation in which any central cold perception was denied (BENZINGER et al., 1961). But this view contrasted with the well-established observation that cooling of the whole body core when the skin was warm regularly evoked strong cold defence reactions in dogs (CHATONNET and TANCHE, 1957a, b; HALLWACHS et al., 1961a). This discrepancy greatly stimulated the search for extracerebral deep body thermoreceptors and it is obvious why this search was concentrated, at that time, on the effector responses to central cooling (BLATTEIS, 1960; BLIGH, 1961; HALLWACHS et al., 1961b; DOWNEY and MOTTRAM, 1962).

HALLWACHS et al. (1961b) who cooled the whole extracerebral body core in anesthetized dogs after thermal isolation of the head were the first to demonstrate definite cold defence reactions in response to deep body cooling at elevated skin and brain temperatures. Subsequent investigations by RAUTENBERG et al. (1963a, b) and by SIMON et al. (1963a) in anesthetized dogs confirmed a central cold sensitivity both in the brain and in the extracerebral body core. These results initiated the systematic search for the extracerebral deep cold sensitive structures which finally led to the spinal canal as a site of specific extrahypothalamic thermosensitivity.

1.1.1. First Evidence for Thermosensitive Structures in the Spinal Canal

The responses of two thermoregulatory effector systems to selective temperature changes within the spinal canal of anesthetized dogs furnished the first evidence for thermal susceptibility with an apparently specific function in a circumscribed region of the extrahypothalamic body core. Fig. 1 firstly demonstrates the activation of metabolic heat production due to shivering which was induced by selective cooling of the spinal canal at elevated core and skin temperatures. Secondly, the course of the skin temperature measured at the paw indicates that skin blood flow was reduced by cold stimulation within the spinal canal (SIMON et al., 1963b).

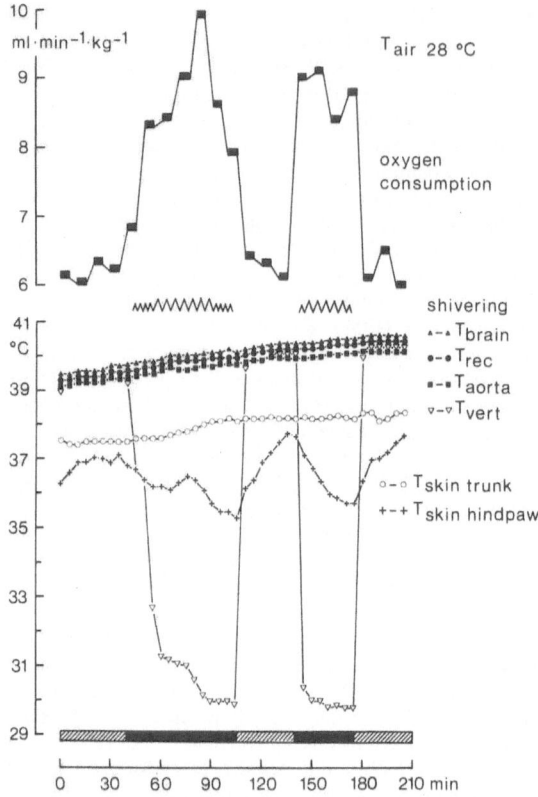

Fig. 1. Selective cooling (black bars) within the lower thoracic and lumbar spinal canal of an anesthetized dog by means of a water perfused peridural thermode at warm ambient conditions. Shivering and increased oxygen consumption are indued by lowering spinal canal temperature in spite of elevated, rising extraspinal core temperatures. The fall of skin temperature at the paw during spinal cooling indicates cutaneous vasoconstriction. (SIMON et al., 1963b)

Further investigations in anesthetized dogs confirmed that selective lowering of the spinal canal temperature regularly evoked cold shivering in the absence of any other central or peripheral cold stimuli (SIMON et al., 1964). RAUTENBERG and SIMON (1964) also demonstrated an appropriate interaction between external and central cold stimuli and cold and warm stimuli applied to the spinal canal. Shivering which had been evoked by peripheral or by general body cooling was further enhanced by spinal canal cooling and was reduced or abolished by spinal canal warming. Subsequently, SIMON et al. (1965) showed that shivering could be regularly induced in unanesthetized dogs by selective cooling within the spinal canal at thermoneutral ambient conditions (Fig. 2).

1.1.2. Methods of Spinal Thermal Stimulation

Perfusion of parts of the spinal subarachnoidal space with artificial liquor of various temperatures was originally employed as the method of thermal stimula-

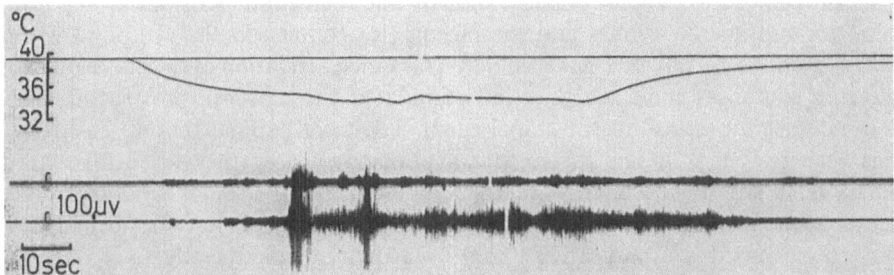

Fig. 2. Electromyograms recorded from the fore and hind legs (lower traces of the recording) of a conscious dog demonstrating shivering induced by selective cooling of the spinal canal at thermoneutral ambient conditions ($T_a = 23°$ C). A peridural thermode extending from the second cervical vertebra to the sacral bone is perfused with water of 22° C. Upper traces: constant rectal and changing spinal canal temperatures. (SIMON et al., 1965)

tion. Later on, double-barrelled, U-shaped thermodes of polyethylene tubing were inserted in the peridural space and were perfused with water of varying tempera-ture. Although greater transverse temperature gradients were produced in the spinal canal by thermode stimulation, this became the standard stimulation method since chronically implanted thermodes were tolerated well and allowed experiments in unanesthetized preparations. Variations of the method concerned the number and the length of the thermodes. Only a few experiments are reported in which metal thermodes instead of the flexible polyethylene tubings were used (GUIEU and HARDY, 1970a, b). Indirect spinal canal cooling and heating by means of thermodes located outside of but adjacent to the spinal canal was performed in a limited number of experiments (BRÜCK and WÜNNENBERG, 1967a, b). Elec-trical, radiofrequency or resistor heating was occasionally employed (BRÜCK and WÜNNENBERG, 1966; WÜNNENBERG and BRÜCK, 1968a, 1970; DUCLAUX et al., 1973).

Due to the anatomy of the spinal canal, quantitative evaluation of stimulus intensity proved to be difficult. Considerable transverse temperature gradients were built up when stimulation was performed by means of thermodes. Especially in bigger species these gradients were found to exceed 10° C if only one thermode was used (CARLISLE and INGRAM, 1973b). These difficulties can partly be overcome by implanting several thermodes, however, their number is limited by the small size of the peridural space. At present, no satisfactory technical solution to the problem of how to uniformly stimulate the spinal canal has been found. In order to estimate average stimulus intensities applied by thermode perfusion at given perfusion temperatures and flow rates, multiple temperature measurements at arbitrary points within the spinal canal were made to permit calculation of "mean spinal cord temperatures" (JESSEN and MAYER, 1971). These average temperatures gave at least reproducible estimations of the stimulus intensity. In other investigations spinal canal temperature was disregarded. Stimulus intensity was simply defined by the temperature of the perfusion fluid.

1.1.3. Location of Spinal Thermosensitive Structures

When the temperature within the spinal canal is experimentally lowered, cold shivering becomes visible in anesthetized dogs within 1–2 min (SIMON et al., 1964) and can be detected by electromyography in conscious dogs within less than

1 min (Simon et al., 1965). Smaller animals like the rabbit (Kosaka et al., 1967; Kosaka and Simon, 1968a) and the pigeon (Rautenberg, 1969) respond within a few seconds to this thermal stimulus. This quick activation of thermoregulatory effector responses indicates that the stimulated thermosensitive structures are located within or close to the spinal canal. The term "spinal thermosensitivity" has been chosen to define those extra-hypothalamic deep thermosensitive structures which are directly affected by the temperature of the spinal canal thermode. Responses evoked by reducing spinal canal temperature below its normal level have been regarded as indicating "cold" sensitivity, while the effects of warming the spinal canal to temperatures above normal have been ascribed to spinal "warm" sensitivity. Both terms will be used in the following discussion in a purely discriptive manner; that is, they are not meant to anticipate a decision about the thermal coefficients of the thermosensitive structures mediating the responses to warming and cooling.

The first observations in dogs, especially those obtained during spinal cooling by means of subarachnoidal perfusion, indicated that spinal thermosensitivity was more or less equally distributed over the whole length of the spinal canal. This corresponded to the oberation that the effects of spinal cold and warm stimulation in dogs (Simon et al., 1964) and pigeons (Rautenberg, 1969) increased with the length of the thermodes. In contrast, Brück and Wünnenberg (1966) found spinal heat sensitivity in guinea pigs to be concentrated in the lower cervical and upper thoracic segments of the spinal canal (Fig. 3). In the pig the cervical part of the spinal canal was also found to be more sensitive to spinal thermal stimulation than the lumbar and sacral sections (Carlisle and Ingram, 1973b).

It is obvious that a part of the central nervous system, the spinal cord, is directly affected by thermal stimulation within the spinal canal. This suggests that the thermally sensitive structures might be localized within the spinal cord itself. The first support for this assumption was obtained from investigations of Meurer et al. (1967). In dogs with chronical bilateral transection of the dorsal roots of the lumbosacral spinal cord; selective cooling of this deafferentiated section of the spinal cord was still effective in evoking cold tremor. Klussmann (1969) who investigated this question with neurophysiological methods found no evidence for activation of afferent dorsal root fibers by thermal stimulation within the spinal canal. Thus, it is most likely that spinal thermosensitivity is based on temperature dependent spinal neurons or neuronal circuits as it is the case with hypothalamic thermosensitivity.

1.2. Specifity and Spectrum of Thermoregulatory Effector Responses Induced by Spinal Thermal Stimulation

Activation and inhibition of shivering by changes of spinal cord temperature as observed in the first explorative studies on spinal thermosensitivity indicated but did not prove its specific function in temperature regulation. Temperature effects on signal transmission in the spinal cord which had repeatedly been described in the past (see Paragraph 1.3.1.) were apparently non-specific, but could have at least partly accounted for the motor reaction of cold induced muscle

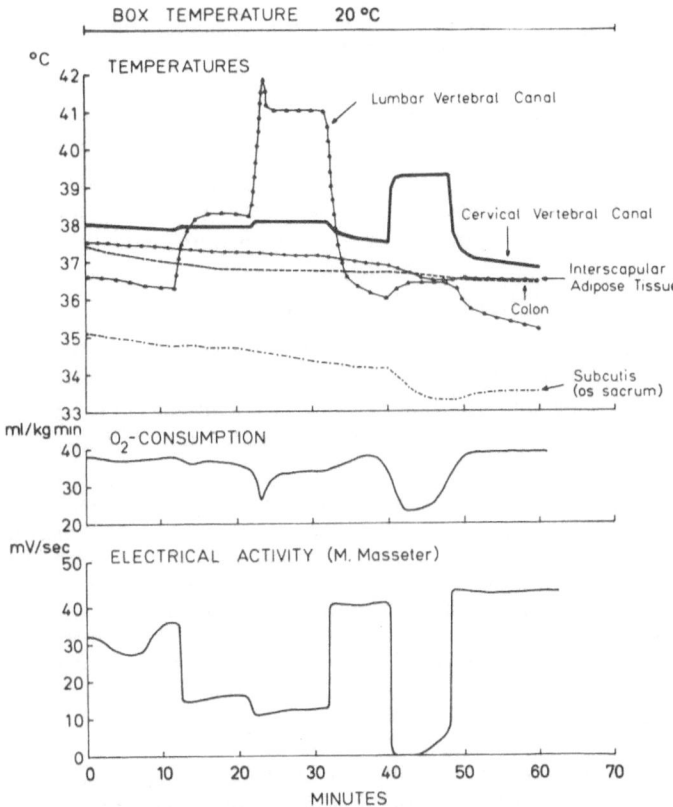

Fig. 3. Oxygen consumption and integrated electrical activity of the electromyogram recorded from the m. masseter of a warm adapted, conscious guinea pig. Shivering due to a moderate peripheral cold stimulus ($T_a = 20°$ C) is almost completely inhibited by radio frequency heating of the cervical spinal cord. The effect of heating the lumbar spinal cord is definitely smaller. (BRÜCK and WÜNNENBERG, 1966, 1970)

tremor. Additional experimental evidence for the specific function of spinal thermosensitive structures was required. The lack of safe neurophysiological criteria for specific neuronal thermosensitivity even in the hypothalamus (EISENMAN, 1972) gave little hope of evaluating the specific nature of spinal thermosensitivity by electrophysiological means. As a consequence, its specific function had to be confirmed by the analysis of effector responses which could be elicited by thermal stimulation of the spinal cord. Investigations which have helped to evaluate the spectrum of thermoregulatory effector responses to spinal cord cooling and warming will subsequently be discussed. Special attention will be paid to the modes of interaction of spinal with hypothalamic or peripheral thermal stimuli and to the comparative analysis of the effector responses. Findings concerning the possible physiological significance of spinal thermosensitivity, for instance the quantitative relations between stimulus intensity and effector responses, will be discussed in a special chapter.

A summary of thermoregulatory effector responses induced by spinal temperature modifications in *unanesthetized animals* is given in the Table (see p. 24 to 26). This listing shows that the influences of spinal thermal stimulation on virtually all known thermoregulatory effectors have been studied by many authors in numerous species. There is ample, unequivocal evidence for appropriate activation or inhibition of most effector systems by changes of spinal cord temperature. Thus, it is generally accepted that the thermosensitive structures of the spinal cord are specifically incorporated in the temperature regulation system.

1.2.1. Shivering and Shivering Thermogenesis

The type of tremor termed shivering is an adequate somatomotor response to cold stress. It is the predominant calorigenic mechanism in many homeothermic species. Analysis by mechanographic and electromyographic recordings (LEVY, 1963; STUART et al., 1963, 1966a–c) have revealed the characteristic features of this cold induced tremor.

Periodic bursts of activity often correlated with the respiratory rhythm are frequently observed both in the mechanogram and electromyogram. These bursts are most marked during the initial stages of shivering. Due to a partial coordination of the discharges from different motor units, a pattern of grouped discharges appears in the electromyogram with frequencies closely related to the mechanographically evaluated tremor rate. Average tremor rate is dependent on body size in a systematic manner as recently shown by electromyographical studies of SPAAN and KLUSSMANN (1970). With some experience in the observation of shivering, this type of tremor can be identified visually and by palpation and can be distinguished with great reliability from muscle jerks and other types of tremor.

Most investigators who observed cold induced tremor during spinal cord cooling have made use of visual and palpatory control and agree that it is real cold shivering that is evoked by the spinal cold stimulus. All reports concerned with cold shivering in conscious animals as influenced by spinal canal temperature are condensed in part 1 of the Table (see p. 24f.) and may be summarized as follows. Shivering can be evoked by spinal cord cooling in all investigated animals, although species differences seem to exist with regard to the responsiveness of the shivering mechanism to the spinal cold stimulus. Shivering induced by external or hypothalamic cold stimulation or by general hypothermia is enhanced by spinal cord cooling and is reduced or abolished by spinal warming. Conversely, shivering which is elicited by spinal cooling may be appropriately influenced by thermal stimuli applied to other peripheral or central temperature sensors. No species has been detected so far in which changes of spinal cord temperature were without any effect on shivering.

Cold shivering induced by peripheral spinal, or other central cold stimuli has been comparatively analyzed by means of electromyography and mechanography in several investigations. Experiments in anesthetized dogs (MEURER et al., 1965) showed that shivering induced by external cooling, by moderate general hypothermia and by selective cooling of the spinal cord exhibited identical tremor frequencies in individual animals. It was further shown that the bursting type of shivering originally termed "frisson réflexe" by CHATONNET and TANCHE (1956) was observed at the initial stages of tremor under all three types of cold stimulation. With increasing intensity and duration tremor activity became continuous. This continuous shivering activity which has been termed "frisson central"

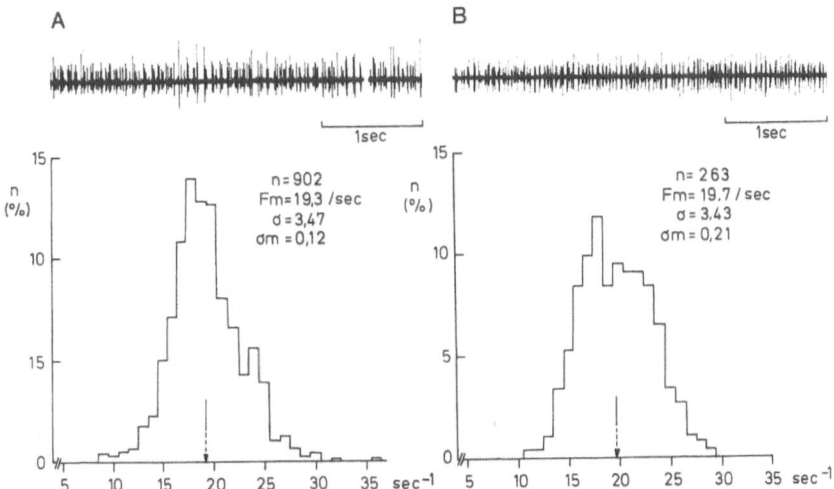

Fig. 4. Grouped discharges in the electromyograms of unanesthetized rabbits obtained from shivering under peripheral cooling (A) and spinal cord cooling (B). Original recordings (above) demonstrate the pattern of the grouped discharges; the histograms (below) show the distribution of grouped discharge frequencies. (KOSAKA and SIMON, 1968b; in: THAUER and SIMON, 1972)

(CHATONNET and TANCHE, 1956) was again observed under all conditions of cold stimulation. Investigations in unanesthetized rabbits with chronically implanted spinal canal thermodes were devoted to a comparative, quantitative analysis of the electromyograms recorded either during peripheral cooling or during spinal cord cooling at warm external conditions (KOSAKA and SIMON, 1968a, b). It was found that the relation between bursting and continuous shivering was similar for both types of cold stimulation. Further, the patterns of grouped discharges in the electromyograms appeared to be identical. Under the same recording conditions the fractions of recording times during which grouped discharges could be identified were 30% during external cooling and 33% during spinal cord cooling. The frequency spectra of the grouped discharges during shivering induced by external cooling and by spinal cord cooling are demonstrated in Fig. 4. Both proved to be closely similar as were the average tremor rates.

The comparative investigations on cold shivering were further extended to the function of the proprioceptive control loop during various types of tremor. These investigations revealed that the effects of peripheral and spinal cold stimulation on the function of the γ-motor-muscle spindle system corresponded to each other in several respects but also showed distinct differences. As shown by Fig. 5, dynamic sensitivity to stretch of primary spindle receptors is considerably reduced during spinal cord cooling. This phenomenon is also found in the tremor induced by injecting Curare into the 3rd cerebral ventricle (KLUSSMANN and HENATSCH, 1969) or when shivering is induced by peripheral cooling (SCHÄFER, 1969; SCHÄFER and SCHÄFER, 1973). Apparently, instability of the proprioceptive servo control system and, consequently, facilitation of shivering results from both spinal and peripheral cold stimulation. The effects of these stimuli differ in that the spontaneous discharges of the primary muscle spindle receptors are permanently

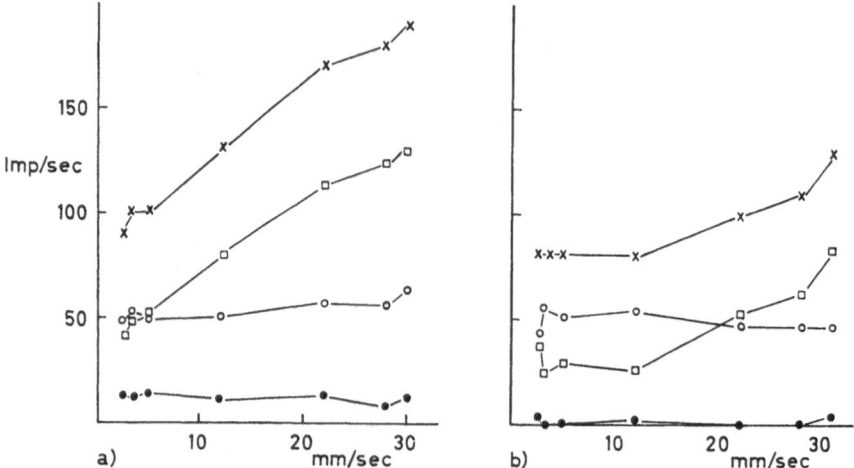

Fig. 5. Effects of spinal cord cooling on the stretch response of a primary muscle spindle receptor from the m. triceps surae of the cat (ramp stretch of 10 mm) as determined from its afferent discharge. Spontaneous activity (●), static response (○), the dynamic component of the response (□), and peak response (×) are plotted against stretch rate a) before and b) during spinal cord cooling of 10 min duration. Notice the reduction of rate sensitivity during cooling, especially at low stretch rates. (Klussmann and Henatsch, 1969)

increased during shivering induced from the periphery, while spinal cord cooling is followed by a transient increase and a subsequent reduction of spindle activity. Consequently the authors have arrived at different conclusions with regard to the changes of γ-motor innervation induced by spinal and peripheral cold stimuli. The prevailing depression of γ-activity during prolonged spinal cooling as observed by Klussmann and Henatsch (1969) is ascribed to local temperature effects on the spinal motor efferents (see Paragraph 1.3.1.). The transient activation of γ-motor activity during the initial stages of spinal cold stimulation might correspond to the increased activity of the dynamic γ-fibers during peripheral cooling which, according to Schäfer and Schäfer (1973), accounted for the changes of muscle spindle function in this situation.

The changes of proprioceptor function associated with peripheral or spinal cold stimulation may facilitate and coordinate the shivering tremor but are apparently not responsible for the generation of the tremor rhythm. This is indicated by the onset of shivering in response to cold stimulation in animals with dorsal root transection (Stuart et al., 1966a; Meurer et al., 1967). Tremor frequency and partial synchronization of motor unit activity are rather determined by an intra-spinal tremor mechanism which probably consists of particular neuronal circuits. Spaan and Klussmann (1970) have contributed evidence to this view by analyzing the possible interrelations between reflex and contraction times and shivering frequencies in various species of different body sizes.

Different effects of spinal and peripheral cold stimulation on extrafusal muscle activity have been described by Stelter et al. (1969). Reflex tension of "fast" muscles was found to be increased at slight to moderate degrees of spinal cooling, reflex tension of "slow" muscles was uniformly depressed. In contrast, peripheral cooling increased reflex tension in both types of muscles. According to the observations of Klussmann (1964, 1969), this differential effect of spinal cooling is due to the different susceptibilities of the various types of α-motoneurons to the local temperature changes. These local temperature effects on somatomotor efferents may modify shivering intensity but appear to have no association with the generation of the specific shivering tremor during spinal cord cooling.

The influences of spinal cord temperature on metabolic heat production associated with cold shivering have been evaluated in numerous investigations by measuring oxygen consumption. The rise of metabolic heat production, which was regularly induced by spinal cooling in all investigated species, was found to be related not only to cooling intensity but also to other thermal factors such as skin and hypothalamic temperatures. For instance, a maximum increase of heat production by 200–300% of the basal metabolic rate could be induced by spinal cord cooling in the conscious dog at cool ambient conditions ($T_a = 18°$ C). In a warm environment ($T_a = 30°$ C) the same animal responded to an equivalent spinal cold stimulus with an increase of less than 100%. The metabolic response to hypothalamic cooling was found to be influenced in the same way by ambient air temperature (JESSEN and MAYER, 1971). In the conscious pig oxygen consumption linearly increased with increasing intensity of spinal cord cooling at cool ambient conditions ($T_a = 15°$ C). In a warm environment ($T_a = 33°$ C) spinal cord cooling had no effect on oxygen consumption (CARLISLE and INGRAM, 1973b). In both species an appropriate interaction has been observed between the spinal cord and the hypothalamus with regard to the control of metabolic heat production. This is demonstrated by the data of Fig. 6 obtained from a conscious dog. The activating effect of simultaneous cooling of both central thermosensitive regions was distinctly greater than cooling of one area alone. On the other hand, the stimulating effect of spinal cord cooling on heat production could be cancelled by hypothalamic heating and vice versa (JESSEN et al., 1968; JESSEN and LUDWIG, 1971; JESSEN and SIMON, 1971). CARLISLE and INGRAM (1973a) have observed the same kind of interaction between spinal cord and hypothalamus in the conscious pig. Apparently, simultaneous, equidirectional temperature displacements of hypothalamus, spinal cord and skin will reinforce each other, while opposing stimuli tend to neutralize each other. This finding strongly indicates a specific kind of interaction between these three thermosensitive regions of the body.

The interaction between spinal and hypothalamic thermal stimuli with respect to metabolic heat production seems to exhibit species differences regarding the relative sensitivities of both central thermosensitive regions. In the guinea pig (BRÜCK and SCHWENNICKE, 1971) and the pigeon (RAUTENBERG et al., 1972) hypothalamic temperature displacements seem to exert little, if any, influence on shivering, while changes of spinal cord temperature are highly effective. As a consequence, little interaction is observed when the temperatures of both regions are simultaneously displaced in the same or opposite directions. An example is given in Fig. 7 in which a conscious pigeon was submitted to spinal cord cooling and to warming and cooling of the brain. Heat production and body heat content were definitely increased by spinal cord cooling while brain cooling was ineffective. Accordingly, and in contrast to the dog (Fig. 6), simultaneous cooling of the brain and the spinal cord did not augment the shivering response in comparison to selective spinal cooling. Also, brain heating during spinal cooling had no inhibiting influence. However, the course of the foot temperature indicates that brain cooling may be effective in reducing skin blood flow of pigeons.

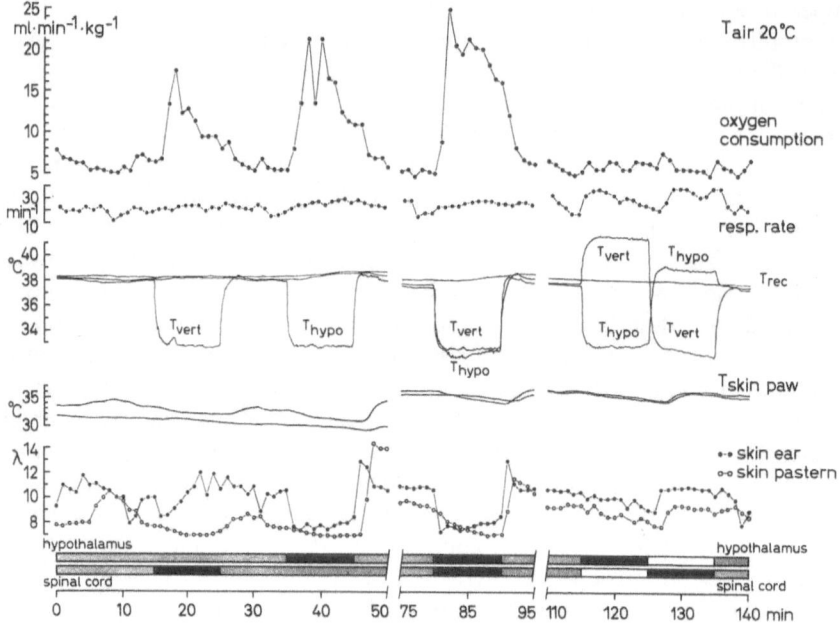

Fig. 6. Selective and simultaneous temperature changes of the spinal cord and hypothalamus in a conscious dog. Ambient air temperature 20° C; effects on oxygen consumption and respiratory rate. Direct recordings of peridural spinal canal temperature (T_{vert}), of hypothalamic temperature (T_{hypo}), of rectal temperature (T_{rec}), and of two paw skin temperatures. The lower ordinate (λ) indicates the changes of heat conductivity of the skin at the ear and the hind leg as determined with the heated thermocouple method. Thermal states of the hypothalamus (upper horizontal bars) and spinal canal (lower horizontal bars); hatched: no stimulation, black: cooling, white: heating. (JESSEN et al., 1968)

Humoral aspects of the shivering response to spinal cooling have recently been investigated by CLOUGH et al. (1973). In shorn sheep spinal cooling led to increased oxygen consumption which was associated with a rise of the free fatty acid blood levels. This cooling effect has to be regarded as a typical humoral response to cold stress, as it is also found during external cold exposure in sheep (BOST and DORLEAC, 1965) and in other species. The fact that β-receptor blocking agents and splanchnicotomy were able to block the increase of the free fatty acid blood levels during spinal cooling as well as under other cooling conditions shows that this response is mediated by the sympathoadrenal system.

The analysis of the role spinal nervous structures play in the generation of cold shivering has not only confirmed the existence of a spinal temperature reception. It has further led to the discovery of *subsidiary spinal control functions in temperature regulation.* In anesthetized dogs with high spinal transections a clearly visible shivering-like muscle tremor can be induced by spinal cord cooling (SIMON et al., 1966). Part I A of Fig. 8 demonstrates the mechanogram of one hind leg and the electromyogram of the dorsal trunk muscles in a spinalized anesthetized dog during spinal cord cooling. Clearly, this cold stimulus led to an unmistakable activation in the electromyogram and to fine tremor movements of the hind leg as visualized by the mechanogram. Part I B shows that the electromyogram was composed of grouped discharges as in normal shivering. However,

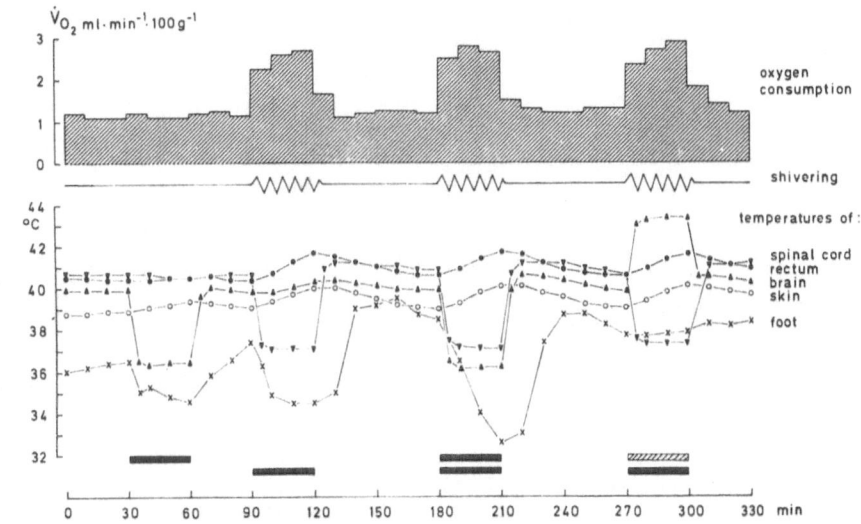

Fig. 7. Cooling of the brain and the spinal cord of a conscious pigeon at warm ambient conditions ($T_a = 29.5°$ C); effects on oxygen consumption and on foot skin temperature. Horizontal bars mark thermal stimulation of the brain (upper horizontal bars) and of spinal cord (lower horizontal bars); black: cooling, hatched: heating. Brain cooling is ineffective in evoking shivering, brain heating is ineffective in suppressing shivering induced by spinal canal cooling. The foot temperature indicates that both brain and spinal cord cooling elicit cutaneous vasoconstriction. (RAUTENBERG et al., 1972)

a comparison of the mechanograms obtained from the same animal under identical recording conditions before (part II) and after (part I A) spinal transection clearly demonstrates that the intensity of the tremor was considerably reduced in the spinalized preparation. Despite this quantitative difference, the patterns of cold induced electromyographical activity in spinal animals were found to be closely similar to those of normal electromyographical shivering. Subsequent investigations in chronically spinalized rabbits have further shown that not only spinal cord cooling but also external cooling could induce increased activity in the electromyogram (KOSAKA and SIMON, 1968a, b). A quantitative analysis of the electromyographical activity induced in spinalized rabbits by spinal and external cooling was carried out in these investigations. The results were compared with those obtained from rabbits with an intact spinal cord. This comparison revealed that the patterns of shivering in normal and spinal animals were closely similar with respect to the frequency distribution of the grouped discharges in the electromyogram. Thus, the suggestion was confirmed that the spinal structures are able to transform a cold stimulus into an appropriate thermoregulatory effector response. This would mean that thermoregulatory control mechanisms, though rudimentary in nature and of a limited efficiency, are localized in the spinal cord. This conclusion has suggested the hypothesis that the evolutionary development of the central control mechanisms for this cold defence response may have originated at the segmental level of the central nervous organization

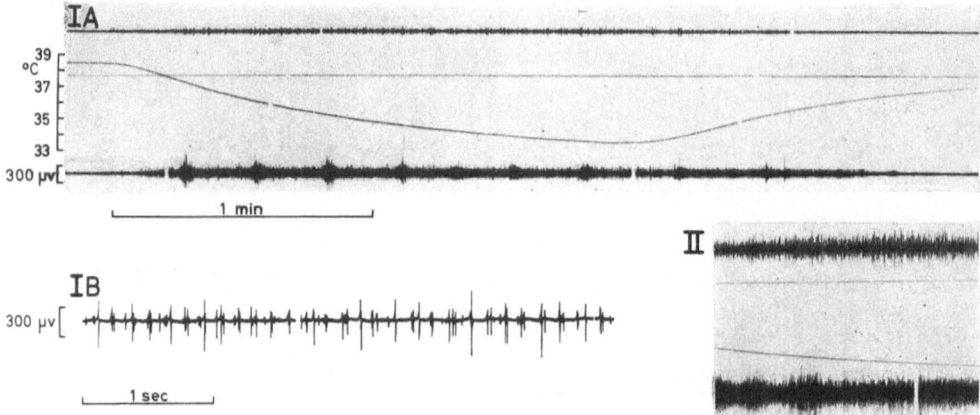

Fig. 8. Shivering recorded in an anesthetized dog with spinal transection at C1/C2. A visible tremor was elicited by spinal cord cooling. Part IA: Upper trace, mechanogram of the hind leg. Middle traces, the constant rectal temperature and the changing spinal canal temperature. Lower trace, electromyogram of the dorsal trunk muscle. Part IB: electromyographical activity recorded at higher paper speed shows grouped discharges. Part II: it shows, for reasons of comparison, the mechanogram, the rectal and spinal canal temperatures and the electromyogram recorded during spinal cooling in the same animal under identical recording conditions before spinal transection. (Simon et al., 1966)

(Simon, 1967). These findings have further confirmed the suggestions of Jung (1941), Birzis and Hemingway (1957), Stuart et al. (1963) and others that the nervous mechanism generating the rhythm of cold shivering is segmentally arranged.

1.2.2. Non-Shivering Thermogenesis (NST)

The influence of spinal thermal stimulation on NST has been investigated up to now only in the new-born or cold adapted guinea pig (Brück and Wünnenberg, 1966, 1967a, b; Brück et al., 1970, 1971). Although no evidence for a direct influence of spinal cord temperature on NST was found in this species (Wünnenberg and Brück, 1968b), it was shown that shivering and non-shivering thermogenesis are linked by the spinal thermosensitive structures in a particular manner. In the guinea pig, NST seems to be governed mainly by two thermal inputs: it is driven by cutaneous cold stimuli and is inhibited by hypothalamic warm stimuli (Brück et al., 1967; Brück and Schwennicke, 1971). Shivering heat production is likewise governed by two thermal inputs: it is activated by cooling of the skin and is inhibited by warming of the spinal cord. An unique mode of "meshing" of the two cold defence mechanisms is established by the topographic relations between the subscapular brown adipose tissue, an important source of heat in NST, and the spinal warm sensitive structures which are concentrated in the lower cervical and upper thoracic segments of the spinal cord (Brück and Wünnenberg, 1970). When NST is induced in a newborn or cold adapted guinea pig by external cooling, the heat liberated in the subscapular brown adipose tissue is transferred via the venous drainage to the lower cervical and upper thoracic spinal canal. Thus, spinal cord temperature is kept elevated above the remaining core temperature and the less economic shivering thermogenesis is effectively suppressed by the spinal warm sensitive structures. This situation is found during the first cooling period demonstrated in Fig. 9. Suppression of cold shivering by NST is particularly effective in cold adapted guinea pigs due to the fact that the spinal threshold temperature for shivering is lower in these animals than in warm adapted guinea pigs (BRÜCK and WÜNNENBERG, 1967b; BRÜCK et al., 1971). Non-shivering thermogenesis is controlled by the sympathoadrenal system. Accordingly, the described kind of "meshing" between the control mechanisms for shivering and NST may be experimentally interrupted by administration of a β-blocking agent which suppresses NST. When this is done in the second cooling period of Fig. 9, cold exposure does not result in a rise of spinal cord temperature but instead in its decrease. As a consequence, shivering ensues with the animal responding to the cold stress like warm adapted guinea pigs in which NST is not developed.

1.2.3. Skin Blood Flow

Adjustment of cutaneous blood circulation is an important thermoregulatory effector mechanism by which heat transfer from the body core to the surface is controlled. Effects of changes in spinal cord temperature on skin blood flow have been observed with different methods in various species. For example skin temperature measurements at the ear pinna, paw or tail at controlled ambient conditions, plethysmography, the heated thermocouple method, and electromagnetic flow measurements have been applied. As shown by part 3 of the Table (see p. 25), all investigators agree that skin blood flow is appropriately influenced by thermal stimulation of the spinal cord; namely, spinal cooling reduces skin blood flow, while spinal warming causes cutaneous vasodilatation. Investigations of JESSEN et al. (1967) demonstrated that the vasodilatating influence of spinal cord warming was able to overcome the vasoconstrictor drive of a strong external cold stimulus. IRIKI (1968) was the first to show that the intensity of the spinal warm stimulus required to elicit vasodilatation in the conscious rabbit's ear depended on the external thermal conditions. While at an ambient temperature of 17° C spinal cord temperature had to be raised by 2.3° C on the average to initiate a response, 0.6° C were sufficient to induce skin vasodilatation at an ambient temperature of 25° C. The interaction of skin and spinal cord temperatures with regard to the control of skin blood flow was further demonstrated by INGRAM and LEGGE (1971) in the conscious pig. Fig. 10 shows that at thermoneutral ambient conditions changes of thermode temperature in the range between 35 and 43° C could change skin blood flow from full vasoconstriction to full vasodilatation. At lower or higher ambient temperatures the effects on skin blood flow of the same spinal temperature displacements were much reduced or absent. Interactions between the signals of cutaneous thermoreceptors and spinal thermal stimuli were also demonstrated in experiments in which the scrotal skin of conscious pigs was selectively heated (INGRAM and LEGGE, 1972b). The strong vaso-

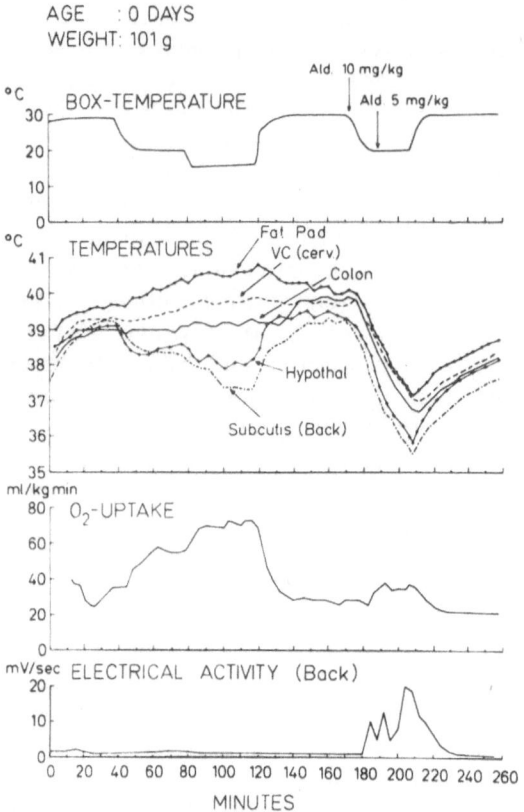

Fig. 9. Oxygen consumption and integrated activity of the electromyogram recorded from the back muscles of a newborn, conscious guinea pig. Non-shivering thermogenesis (rise of oxygen consumption but no electrical activity) is induced by lowering ambient (box) temperature and by a consecutive fall of hypothalamic temperature. The temperature of the cervical spinal canal (vc) is kept elevated due to the heat supply from the subscapular brown adipose tissue (fat pad). After non-shivering thermogenesis is suppressed by β-receptor blockade with Alderlin (Ald.), spinal canal temperature decreases together with the other core temperatures and oxygen consumption is increased by shivering. (Brück and Wünnenberg, 1966, 1970)

dilatatory effect of scrotal heating could be suppressed by simultaneous spinal cord cooling. In this respect the spinal cold stimulus was found to be equivalent to hypothalamic cold stimulation. Further observations of Ingram and Legge (1971) confirmed that hypothalamic and spinal thermal stimuli interacted appropriately in determining skin blood flow. On the whole, their findings correspond to the view that, as with shivering, equidirectional temperature displacements of skin, hypothalamus and spinal cord tend to reinforce each other with regard to the response of the skin vessels, while opposing temperature changes may result in neutralisation.

In order to answer the question of whether or not the response of skin blood flow to spinal thermal stimulation exhibited specific features common to thermoregulatory responses induced by stimulation of other sensors, the cardiovascular

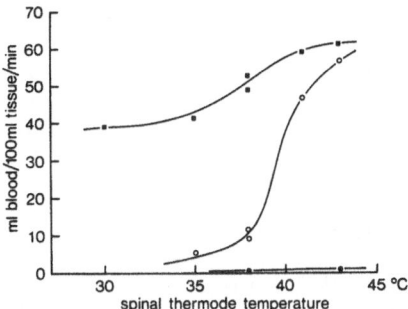

Fig. 10. The influence of spinal thermal stimulation (abscissa: thermode temperature) on skin (tail) blood flow (ordinate) of the conscious pig at different ambient air temperatures. Skin blood flow is varied over its full range by spinal temperature displacements only at $T_a = 25°$ C (o). At $T_a = 30°$ C (■) the skin vessels remain dilated, while at $T_a = 20°$ C (●) vasoconstriction persists. (INGRAM and LEGGE, 1971)

reactions induced by spinal, hypothalamic and peripheral thermal stimulation were comparatively analyzed. For this purpose, blood flow in peripheral and central vascular regions, cardiac function and regional sympathetic activity were studied under experimental conditions in which secondary effects on the circulation due to shivering or panting had been excluded. These investigations revealed a surprising degree of correspondence between the patterns of regional cardiovascular adjustments induced by heat stimulation and, respectively, cold stimulation of each thermosensitive area.

Redistribution of blood flow during external cooling as indicated by a simultaneous decrease in flow to the skin and increase in flow to central vascular regions had already been demonstrated several decades ago by REIN (1931) in the dog. GRAYSON (1949) had also described opposite blood flow adjustments resulting in an increase of skin blood flow and reduction of intestinal blood flow in response to external heating in man. The same types of regional blood flow adjustments could be demonstrated during heating and cooling of the hypothalamus or spinal cord in anesthetized and conscious dogs by means of electromagnetic flow measurements in arteries supplying cutaneous and intestinal vascular regions (KULLMANN et al., 1970; SCHÖNUNG et al., 1971a, b). These differential changes of cutaneous and intestinal blood flows strongly suggested regionally opposite modifications in vasoconstrictor tone as the underlying nervous mechanism. This was confirmed by experiments in anesthetized rabbits in which the discharges of sympathetic efferents supplying the skin and the intestine were simultaneously recorded. Cooling of the skin, spinal cord and, less distinctly, of the hypothalamus was followed by an increase of vasoconstrictor activity in the skin, while intestinal sympathetic activity was reduced. Heating of these regions evoked the opposite pattern of sympathetic responses: cutaneous vasoconstrictor activity was reduced, while intestinal sympathetic activity was increased (WALTHER et al., 1970; IRIKI et al., 1971b; RIEDEL et al., 1972).

The investigations of regional sympathetic responses to peripheral, spinal and hypothalamic thermal stimulation have demonstrated that particular patterns of differential vasomotor responses are involved in thermally induced adjustments of skin blood flow. These patterns are basically identical, regardless of the thermosensitive area which is being stimulated. This finding supports the view that the spinal thermosensitive structures contribute a specific sensory input into the temperature regulation system as do the cutaneous and hypothalamic thermo-receptors.

Apart from the importance of these findings for the assessment of the specific character of spinal thermosensitivity, these studies are of general interest because they have demonstrated for the first time bidirectional, opposing changes of regional activity in the sympathetic outflow controlling the cardiovascular system (Simon, 1971). Thus, a principle of nervous control of regional cardiovascular functions has definitely been established which had been a matter of discussion since Dastre and Morat (1884). Its validity has meanwhile been confirmed by our own and by independent investigations under various conditions of natural stimulations (Horey-sek and Jänig, 1971; Iriki et al., 1971a, c, 1972; Iriki and Simon, 1973; Karim et al., 1971, 1972 Wallin et al., 1971; Delius et al., 1972; Kullmann and Junk, 1973).

Analysis of skin blood flow and of the underlying sympathetic responses during spinal cord cooling in chronically spinalized animals has contributed further support to the idea that *subsidiary spinal thermoregulatory mechanisms* may become active if the spinal cord is deprived from supraspinal influences (Simon, 1968). Walther et al. (1971b) showed that chronically spinalized, conscious dogs were able to respond with cutaneous vasoconstriction to selective cooling of the spinal cord and, though less regularly, also to peripheral cold. Fig. 11 demonstrates that the spinal cold stimulus was very effective in reducing skin blood flow in the spinal animal. Walther et al. (1971a) analyzed the cold induced vasomotor response pattern in spinalized rabbits and found opposing changes of activity in cutaneous and intestinal sympathetic efferents in response to spinal cord cooling a few days after spinal transection. This pattern corresponded to the differential vasomotor response to cold of intact animals thereby confirming the thermoregulatory character of the cold induced vasomotor response of the spinal animal. This finding offers an explanation for the discovery of Thauer (1935) that rabbits with high spinal transection regain a limited resistance against cold.

1.2.4. Panting, Thermal Tachypnea, Respiratory Evaporative Heat Loss

In many species increased ventilation is an important means by which evaporative cooling of the body is achieved. In dogs, cattle, birds and other animals in which this effector mechanism is highly developed, increased dead space ventilation is the main way in which respiratory evaporative heat loss is augmented. These animals exhibit a specific kind of rapid, shallow, open mouth breathing commonly termed thermal panting which is easily recognized as a thermoregulatory response. In other species plain tachypnea is the respiratory response to heat stress. Some animals respond to a severe heat load with a less frequent and deeper breathing, the so called "second phase panting", by which respiratory heat loss is further augmented at the expense of a sometimes excessive respiratory alcalosis.

As shown by part 4 of the Table (see p. 25f.), there is so far no exception from the rule that spinal cord heating is able to elicit tachypnea or thermal panting. The thermoregulatory character of this response follows from the typical features of panting which have been found during spinal heating in dogs, birds and oxen. The respiratory responses to spinal cord heating are, as a rule, preceded or accompanied by inhibition of heat accumulating mechanisms and by activation of other heat loss mechanisms such as skin vasodilatation and sweating. This

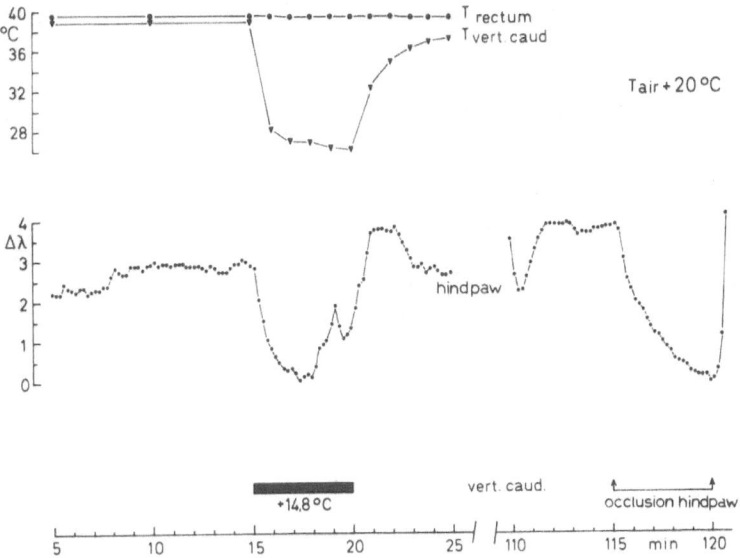

Fig. 11. Changes of cutaneous heat conductivity (λ) determined with the heated thermocouple method at the hindpaw during spinal cord cooling (black bar, with temperature of thermode perfusion) below the level of chronical spinal transection in a dog (32 days after transection). Paw exposed to neutral ambient air temperature. The change of heat conductivity during arterial occlusion (zero blood flow) is shown on the right side. (WALTHER et al., 1971 b)

holds true also for animals in which thermal tachypnea is a less prominent effector mechanism.

In various species the interactions between thermal stimuli applied to the spinal cord and thermal stimulation of other thermosensitive areas of the body core and shell (hypothalamus, abdomen, whole body skin, scrotum) were investigated with regard to the panting response. On the whole, the results were less homogeneous than those reported for metabolic heat production and skin blood flow. Observations of JESSEN et al. (1967) and of JESSEN (1967) showed that panting induced by spinal heating was only transitory at low ambient temperature but was sustained at thermoneutral or warm conditions. In the pigeon panting induced by external heating was effectively inhibited by spinal cooling (RAUTENBERG, 1969). Tachypnea elicited by scrotal heating in the pig was inhibited not only by hypothalamic cooling but also by cooling of the spinal cord (INGRAM and LEGGE, 1972b). INGRAM and LEGGE (1972a) further showed that heat stimuli applied to the whole body skin, the hypothalamus and the spinal cord reinforced each other in eliciting tachypnea in the pig. In the same species tachypnea induced by hypothalamic heating was found to be suppressed by spinal cooling and vice versa. RIEDEL et al. (1973) showed that tachypnea induced by intra-abdominal heating in the rabbit was further augmented by additional spinal cord heating. In the ox, second phase panting could be induced by spinal heating at warm ambient conditions (MCLEAN et al., 1970). All these results fit in with the kind

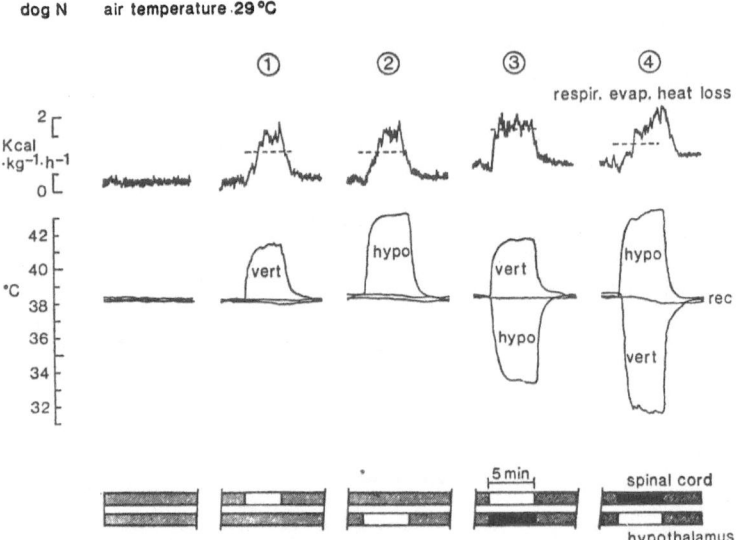

Fig. 12. Selective and simultaneous opposite temperature changes of the spinal cord and the hypothalamus in a conscious dog. Ambient air temperature 29° C; effects on respiratory evaporative heat loss. Direct recordings of peridural spinal canal temperature (vert), of rectal temperature (rec) and of hypothalamic temperature (hypo). Thermal states of the spinal canal (upper horizontal bars) and hypothalamus (lower horizontal bars); hatched: no stimulation, black: cooling, white: heating. Spinal cord heating (1), hypothalamic heating (2), simultaneous spinal cord heating and hypothalamic cooling (3), simultaneous hypothalamic heating and spinal cord cooling (4). (Jessen and Simon, 1971)

of interaction between different thermosensitive areas which was observed in the investigations on metabolic and skin blood flow responses to thermal stimulation.

Several exceptions from this rule were also observed, however. Guieu and Hardy (1970a) were not able to inhibit panting by spinal cord cooling in the rabbit if it had been evoked by hypothalamic heating. Jessen and Simon (1971) confirmed this observation in conscious dogs but showed in addition that panting induced by strong spinal heating could likewise not be inhibited by even intensive hypothalamic cooling (see Fig. 12). In the pigeon, panting induced by spinal heating was not influenced by changes of brain temperature (Rautenberg et al., 1972). Ingram and Legge (1972a) found that panting induced by external heating of the whole body in pigs could not be reduced by either spinal cord or hypothalamic cooling but only by simultaneous cooling of both areas. In the ox and goat, spinal cooling had only little effect on panting in contrast to hypothalamic cooling, but spinal heating was very effective in evoking panting (Jessen et al., 1972; Jessen and Clough, 1973a, b). Taken together, the panting responses to simultaneously applied heat and cold stimuli indicate, in several species, a kind of signal interaction which differs from that observed in the control of shivering and skin blood flow. These deviations from the general rule during panting suggest a preponderance of central heat signals over central cold signals in the control of this effector response.

1.2.5. Sweating

Evaporative heat loss from the skin which is a most important effector mechanism in man has been investigated only in the ox with regard to the influences arising from spinal thermal stimulation (see part 5 of the Table, p. 26). HALES andJESSEN (1969) could demonstrate that spinal cord heating was able to increase water loss from the skin as well as to activate other heat dissipating mechanisms such as skin vasodilatation and thermal panting. In another study, intensive spinal cord heating was found to be a strong enough stimulus to maintain elevated heat loss through all pathways of heat dissipation in spite of peripheral cold stimulation and a considerable degree of general hypothermia (JESSEN et al., 1972).

1.2.6. Piloerection

This autonomic thermoregulatory response which increases the content of still air in the insulating body coat, serves as a heat retaining mechanism in birds and furred mammals. Piloerection as influenced by spinal cord temperature was first described by RAUTENBERG (1969) in pigeons. Fig. 13 demonstrates fluffing in a pigeon in which the spinal cord was cooled at warm ambient conditions. Conversely, fluffing induced by external cold was reduced by spinal heating. Besides this autonomic response, the bird showed a postural response to the spinal cold stimulus by squatting down to cover its legs with its plumage. As further shown by part 6 of the Table (see p. 26), piloerection induced by spinal cord cooling has recently been observed also in the ox, dog, monkey and rat.

1.2.7. Behavior

The behavioral aspects of thermoregulation which play a predominant role in human adaptation to adverse environments have gained increasing importance in the experimental research of temperature regulation in animals during the last few years. The investigations in this field, in particular those concerned with operant behavior and the search for a preferred ambient temperature, have recently been reviewed by CORBIT (1970), CABANAC (1972) and by HAMMEL et al. (1973). Feeding and drinking behavior is also closely related to body temperature (ANDERSSON et al., 1963b; Hamilton, 1967).

Several recent reports have been devoted to the question, whether spinal thermosensitive structures are involved in the control of thermoregulatory behavior (see part 7 of the Table, p. 26). The first description of a behavioral response by *postural adjustment* was given by RAUTENBERG (1967) as demonstrated in Fig. 13. CARLISLE and INGRAM (1973b) have recently reported that spinal cord cooling evokes a postural response in the pig which is identical with that induced by external cooling. With regard to this type of thermoregulatory behavior, the spinal cold stimulus was found to be more effective than a hypothalamic cold stimulus.

In several amphibian species the influence of deep body temperature on the search behavior for the *preferred ambient temperature* was demonstrated by CABANAC and JEDDI (1971). According to DUCLAUX et al. (1973), warm sensitivity

Fig. 13. Changes in the state of piloerection induced by thermal stimulation of the spinal cord in an unanesthetized pigeon. Left side: pigeon at warm ambient air temperature before (above) and during (below) spinal cord cooling. Right side: pigeon at cold ambient air temperature before (above) and during (below) spinal cord heating. Note the change of posture during spinal canal cooling. (RAUTENBERG, 1967; in: THAUER and SIMON, 1972)

of the spinal cord participates in guiding these ectothermic animals towards their external thermopreferendum. This function of the spinal thermosensitive structures corresponds to the influence of the hypothalamic thermoreceptors on the thermoregulatory escape response of lizards (MYRHE and HAMMEL, 1969) and fishes (CRAWSHAW and HAMMEL, 1971).

The influence of changes in spinal cord temperature on *operant behavior* has been investigated in dogs (CORMARÈCHE-LEYDIER and CABANAC, 1973) and in pigs (CARLISLE and INGRAM, 1973b). CORMARÈCHE-LEYDIER and CABANAC (1973) trained dogs to ask for either cold air or radiant heat by interrupting different light beams. Heating the spinal cord at warm ambient conditions increased the animals' demand for cold air. At cold ambient conditions, the rate at which radiant heat was required was reduced by spinal heat stimulation, while the animals' demand for cold air was increased (Fig. 14). These operant responses apparently corresponded to those induced by hypothalamic heating in rats (CARLISLE, 1966; MURGATROYD and HARDY, 1970) and in monkeys (ADAIR, 1970; ADAIR et al., 1970; GALE et al., 1970). However, spinal cord cooling in the dog had no significant influence on operant behavior in contrast to hypothalamic cooling, as

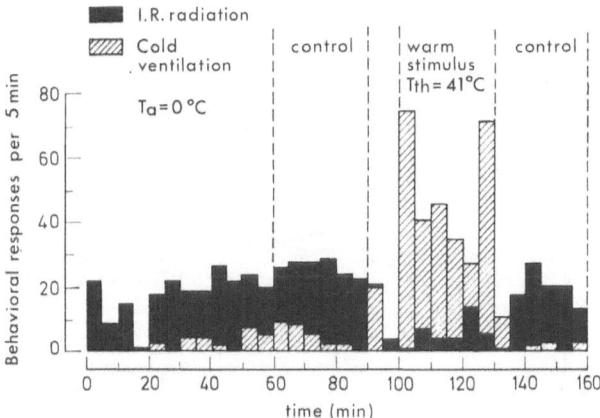

Fig. 14. Operant behavioral thermoregulation of a conscious dog at cold ambient air temperature ($T_a = 0°$ C). Black vertical bars: demands for external infrared heating (2 sec), hatched vertical bars: demands for cold air ventilation (2 sec); ordinate: number of demands per 5 min, abscissa: time in min. During heat stimulation of the spinal cord (perfusion temperature of the spinal canal thermode: T_{th}), the demand for external heating is reduced, while the demand for cold air is considerably increased. (CORMARÈCHE-LEYDIER and CABANAC, 1973)

shown in the investigations in monkeys (ADAIR et al., 1970; GALE et al., 1970) and rats (SATINOFF, 1964). On the other hand, CARLISLE and INGRAM (1973b) were able to demonstrate operant behavior in the pig in response to both elevating and lowering spinal cord temperature. At cool to cold ambient air temperatures, spinal cord cooling increased and warming decreased the rate at which heat reinforcement was required by means of bar pressing. These responses corresponded to the operant behavior of the pigs induced by hypothalamic cooling and warming (BALDWIN and INGRAM, 1967), but appeared to be less pronounced or transitory. If opposing displacements of spinal and hypothalamic temperatures were simultaneously performed in the same animal, its operant behavior was determined by hypothalamic temperature.

Investigations on *feeding behavior* as influenced by spinal cord temperature in dogs (SPECTOR and CORMARÈCHE, 1971) and rats (LIN et al., 1972) have shown that spinal heating reduces food intake in both species. Spinal cooling has no significant effect on feeding behavior of dogs but clearly stimulates food intake in the rat. Fig. 15 demonstrates that long-term thermal stimulation of the spinal cord in rats not only influences food intake but also leads to considerable opposing changes of extraspinal core temperature due to activation of autonomic thermoregulatory mechanisms.

The interrelation between temperature regulation and food intake is as yet little understood, because contradictory reports exist on the role of the hypothalamic temperature sensors in the regulation of food intake (ANDERSSON and LARSSON, 1961; INGRAM, 1968; SPECTOR et al., 1968; HAMILTON and CIACCIA, 1971). It is interesting to note that the effects of spinal thermal stimulation on food intake in rats approximate those induced by peripheral thermal stimulation but oppose those induced by hypothalamic thermal stimulation.

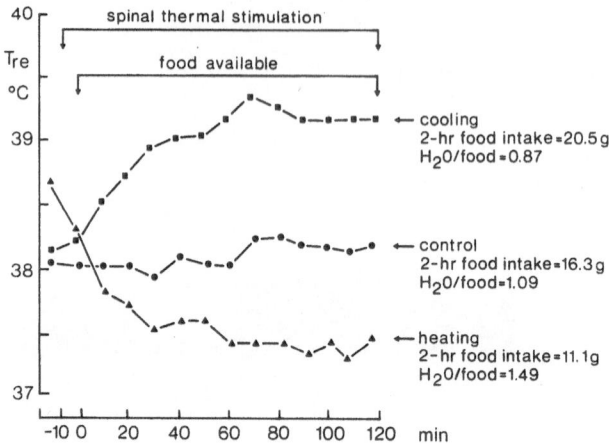

Fig. 15. Effects of heating and cooling of the spinal cord (2 hrs duration) on rectal temperature, food intake and water/food ratio in 4 freely moving rats ($T_a = 25°$ C). Each point in both experimental and control groups represents an average of four animals. (Lin et al., 1972)

Table. Activation or inhibition of thermoregulatory effectors in conscious animals by changes of spinal canal temperature

$+$: activation; $-$: inhibition; $\emptyset$: no effect; ( ): weak or inconsistent effects; [ ] no interaction

Species	Spinal cooling	Spinal heating	Interaction with other thermosensitive areas	References
Part 1: Shivering, shivering heat production				
Dog	$+$			Simon et al. 1965
Dog		$-$	skin	Jessen et al., 1967
Dog	$+$		hypothalamus	Jessen et al., 1968
Dog	$+$	$-$		Jessen, 1971a
Dog	$+$		hypothalamus	Jessen, 1971b
Dog	$+$		skin	Jessen and Mayer, 1971
Dog	$+$		hypothalamus	Jessen and Ludwig, 1971
Dog	$+$	$-$	hypothalamus	Jessen and Simon, 1971
Dog	$+$			Spector and Cormarèche, 1971
Dog	$+$			Cormarèche-Leydier and Cabanac, 1973
Guinea pig		$-$	skin	Brück and Wünnenberg, 1966
Guinea pig	$+$	$-$	skin	Brück and Wünnenberg, 1967a
Guinea pig	$+$	$-$	skin	Brück and Wünnenberg, 1967b
Guinea pig	$+$	$-$	skin, [hypothalamus]	Brück and Wünnenberg, 1970
Guinea pig		$-$	skin	Wünnenberg and Brück, 1968a
Guinea pig		$-$	skin, [hypothalamus]	Wünnenberg and Brück, 1970
Pigeon	$+$	$-$	skin	Rautenberg, 1969
Pigeon	$+$	$-$	skin	Rautenberg, 1971
Pigeon	$+$	$-$	skin, [diencephalon]	Rautenberg et al., 1972
Pig	$+$	$-$	skin, hypothalamus	Carlisle and Ingram, 1973a
Monkey	$+$	$-$		Chai and Lin, 1972
Rat	$+$			Lin et al., 1972

Table (continued)

Species	Spinal cooling	Spinal heating	Interaction with other thermosensitive areas	References
Ox	(+)	—	skin, extraspinal core	JESSEN et al., 1972
Rabbit	+			KOSAKA and SIMON, 1968a
Sheep	+		skin	CLOUGH et al., 1973
Goat	(+)	—	skin	JESSEN and CLOUGH, 1973b

Part 2: Non-shivering thermogenesis

Guinea pig		∅	[skin, hypothalamus]	BRÜCK and WÜNNENBERG, 1970

Part 3: Skin blood flow, sensible heat loss

Dog		+		JESSEN et al., 1967
Dog	—	+	hypothalamus	JESSEN et al., 1968
Dog	—	+		SCHÖNUNG et al., 1971a
Dog	—	+		CORMARÈCHE-LEYDIER and CABANAC, 1973
Rabbit		+	skin, extraspinal core	IRIKI, 1968
Rabbit	—	+	hypothalamus	GUIEU and HARDY, 1970a
Rabbit		+		RIEDEL et al., 1973
Pigeon	—	+	skin	RAUTENBERG, 1969
Pigeon	—	+	diencephalon	RAUTENBERG et al., 1972
Pig	—	+	skin, hypothalamus	INGRAM and LEGGE, 1971
Pig	—	+	scrotal skin	INGRAM and LEGGE, 1972b
Ox		+		HALES and JESSEN, 1969
Ox	(—)	+		MCLEAN et al., 1970
Monkey	—	+		CHAI and LIN, 1972
Rat	—	+		LIN et al., 1972

Part 4: Panting, thermal tachypnea, respiratory evaporative heat loss

Dog		+	skin, extraspinal core	JESSEN et al., 1967
Dog		+	extraspinal core	JESSEN, 1967
Dog		+	hypothalamus	JESSEN and LUDWIG, 1971
Dog		+	skin	JESSEN and MAYER, 1971
Dog	∅		[hypothalamus]	JESSEN and SIMON, 1971
Dog		+		JESSEN, 1971a
Dog		+		SPECTOR and CORMARÈCHE, 1971
Dog		+		CORMARÈCHE-LEYDIER and CABANAC, 1973
Guinea pig		+	skin, hypothalamus	BRÜCK et al., 1971
Guinea pig		+	skin, hypothalamus	HERRMANN, 1972
Pigeon	—	+	skin	RAUTENBERG, 1969
Pigeon	—	+	skin, [diencephalon]	RAUTENBERG et al., 1972
Rabbit		+		KOSAKA et al., 1969b
Rabbit	(—)	+	hypothalamus	GUIEU and HARDY, 1970a
Rabbit		+	abdomen	RIEDEL et al., 1973
Pig	—	+	hypothalamus, skin	INGRAM and LEGGE, 1972a
Pig	—	+	scrotal skin	INGRAM and LEGGE, 1972b

Table (continued)

Species	Spinal cooling	Spinal heating	Interaction with other thermosensitive areas	References
Ox		+		Hales and Jessen, 1969
Ox	(−)	+		McLean et al., 1970
Monkey	−	+		Chai and Lin, 1972
Rat	−	+		Lin et al., 1972
Goat	∅	+	skin	Jessen and Clough, 1973b

Part 5: Sweating, cutaneous evaporative heat loss

Ox		+		Hales and Jessen, 1969
Ox	∅	+		McLean et al., 1970

Part 6: Piloerection

Pigeon	+	−	skin	Rautenberg, 1967
Ox	+			McLean et al., 1970
Monkey	+			Chai and Lin, 1972
Rat	+			Lin et al., 1972
Dog	+			Cormarèche-Leydier and Cabanac, 1973

Part 7: Behavior; (P): postural against cold; (OpH): operant for heat reinforcement; (OpC): operant for cold reinforcement; (T_{pref}): search for thermopreferendum; (F): food intake

Pigeon (P)	+			Rautenberg, 1967, 1969
Pig (P)	+			Carlisle and Ingram, 1973b
Pig (OpH)	+	(−)	hypothalamus	Carlisle and Ingram, 1973b
Dog (OpH)	∅	−		Cormarèche-Leydier and Cabanac, 1973
Dog (OpC)	∅	+		Cormarèche-Leydier and Cabanac, 1973
Dog (F)	∅	−		Spector and Cormarèche, 1971
Rat (F)	+	−		Lin et al., 1972
Frog (T_{pref})		+		Duclaux et al., 1973

1.3. Neuronal Correlates of Spinal Thermosensitivity

It is well established by the numerous investigations discussed in the preceding paragraphs that the centers or pathways of the temperature regulation system must be specifically affected by the temperature dependent discharges of spinal neurons. To date, no evidence has been obtained for the conduction of signals over dorsal roots which could have arisen from thermoreceptor-like structures at the surface of or in close proximity to the spinal cord (Klussmann, 1969). Therefore, it seems likely that spinal thermosensitivity is due to neurons exhibiting high positive or negative temperature coefficients of their discharge rates as it is the case with the thermosensitive neurons discovered in the brain stem. This specific spinal thermosensitivity might be based on the specific thermoreceptive properties of nervous structures. It might, however, be effected as well by a particular mode of interaction between neurons exhibiting a basically non-specific thermal susceptibility.

With regard to the methods applied in the analysis of temperature dependent neuronal activity, the investigations on the preoptic and anterior hypothalamic region seem to have paved the way for the elucidation of neuronal correlates of thermosensitivity in other parts of the central nervous system. The applications of electrophysiological and neuropharmacological methods in the analysis of thermoregulatory functions of the brain stem have furnished essential contributions to the pool of data on which the current neuronal models of temperature regulation have been based.

Electrophysiological methods were introduced in the investigation of hypothalamic thermoregulatory functions by NAKAYAMA et al. (1961, 1963). They have been widely adopted by many investigators (for References see HELLON, 1970a, 1972b; EISENMAN, 1972; HARDY, 1972b) and have been extended to the evaluation not only of local temperature effects but also of influences of remote thermal stimuli on single neurons in the hypothalamus and in other parts of the brain stem (MURAKAMI et al., 1967; WIT and WANG, 1968a; CABANAC and HARDY, 1969; NAKAYAMA and HARDY, 1969; EDINGER and EISENMAN, 1970; HELLON, 1970a, b, 1972a; GUIEU and HARDY, 1970b, 1971; NUTIK, 1971, 1973a, b; WÜNNENBERG and HARDY, 1972; BOULANT and HARDY, 1972). This work has revealed a great variety of temperature dependent responses to changes of local and remote temperatures in the investigated hypothalamic and brain stem neurons. This variety of responses is difficult to incorporate in neuronal models of temperature regulation without loosing lucidity (BLIGH, 1972; Hardy, 1972b). In 1964 the neuropharmacological approach to the analysis of hypothalamic functions in temperature regulation was initiated by FELDBERG and MYERS (1964) and has been widely applied (for References see FELDBERG, 1970; MYERS, 1971; HELLON, 1972c). Several biogenic amines have been recognized as possible transmitters. Although the interpretation of the experimental results is still limited by species differences, several models of temperature regulation have been developed on the basis of these findings (MYERS, 1971; BLIGH, 1972; ZEISBERGER and BRÜCK, 1973). In addition, effects of anesthetics and of pyrogenic and antipyretic substances have been analyzed (CUNNINGHAM et al., 1967; EISENMAN and JACKSON, 1967; MURAKAMI et al., 1967; CABANAC et al., 1968; WIT and WANG, 1968b; EISENMAN, 1969, 1972; BECKMAN and EISENMAN, 1970; BECKMAN and AMINI-KORMI, 1972; HORI and NAKAYAMA, 1973; NAKAYAMA and HORI, 1973). All these various properties of thermosensitive brain stem neurons such as linearity of response to local temperature changes, sensitivity to biogenic amines, pyrogenic substances, antipyretics and anesthetics, and susceptibility to remote, thermal and non-thermal stimuli have been taken into consideration as possible criteria to discriminate between primary sensory, integrating, or efferent functions. However, a discourageing complexity of neuronal responses has still to be encountered in attempting to elucidate in this way the nervous mechanisms of temperature regulation (HARDY and GUIEU, 1971; HARDY, 1972b; WÜNNENBERG and HARDY, 1972). The discovery of highly thermosensitive neurons in brain areas not involved in temperature perception (BARKER and CARPENTER, 1970) has demonstrated that electrophysiological criteria alone do not as yet suffice to unequivocally characterize the thermoregulatory functions of central neurons, a view that was until recently held even for the much better defined cutaneous thermoreceptors (HENSEL, 1970). With regard to the thermosensitive brain stem neurons one is left with the dilemma of recording from an unknown structure with an unknown function (HENSEL, 1973). Considering these difficulties EISENMAN (1972) has arrived at the conclusion that the location of central thermosensitive neurons in a region from which thermoregulatory reactions are elicited is still a most important criterion, if they are to be classified as specific.

In view of the difficulties opposing the interpretation of the data concerning the thermosensitivity of brain stem neurons it has recently been pointed out by

Hardy (1973) that "one of the key problems in understanding the relationship of the unit activity to thermoregulation has been the separation of the *afferent* or input signal from the *efferent* or output signal" and that "it cannot be said that this problem is solved or is near solution ...". Compared with the brain stem, investigations on neuronal thermosensitivity at the spinal level would offer the advantage that afferent and efferent neurons could be more easily distinguished by anatomical and neurophysiological criteria. Thus, it would be possible to decide whether a spinal thermosensitive neuron would be able, in principle, to function as an input channel for supraspinal thermoregulatory centers or whether it would possibly exert some influence on efferent spinal functions involved in the control of thermoregulatory effectors.

The investigation of spinal thermoregulatory functions with electrophysiological means was initiated in 1964 by Klussmann. However, his experimental approach, the analysis of temperature effects on spinal motoneurons, differed from that prevailing in the research on hyopthalamic neuronal functions in temperature regulation. This had been due to the fact that the discovery of spinal thermosensitivity had originally resulted from observations concerning the effects of spinal cooling on the shivering mechanism. These first observations had immediately raised the question what role the long known thermal effects on spinal reflex function might play in this conncetion and had, accordingly, given priority to experiments which contributed answers to this question. Only after the specific effects of spinal cold and heat stimulation on all thermoregulatory effectors had been established, did the input function of spinal thermosensitive structures become the subject of electrophysiological investigations.

1.3.1. Temperature Effects on Efferent Spinal Neurons

Important effector mechanisms of temperature regulation including shivering, non-shivering thermogenesis, skin blood flow, and sweating are predominantly or exclusively controlled by descending spinal pathways and by spinal efferent neurons. Since nervous functions should, in general, follow Arrhenius' law, thermoregulatory effector responses might be influenced locally by temperature changes of the spinal cord even in the absence of specific thermoreceptive neurons.

Temperature effects on signal transmission at the spinal level have, in fact, been known for a long time (Barron and Matthews, 1935, 1938; Tönnies, 1939; Grundfest, 1941; Skoglund and Uvnäs, 1943; Koizumi et al., 1954, 1960; Brooks et al., 1955; Suda et al., 1957). Hyperresponsiveness of motor reflex arcs as described by Brooks and associates is one of those temperature effects which might be involved in the generation and maintenance of shivering during spinal cord cooling.

Klussmann and coworkers (Klussmann, 1964, 1969; Klussmann and Pierau, 1972) have investigated the role of these local temperature effects on the shivering response. Single fiber recordings from ventral and dorsal roots in anesthetized cats revealed that the spinal motoneurons were differentially activated by lowering spinal cord temperature. As shown by Fig. 16, spontaneously firing or silent γ-motoneurons were activated by slight decreases of spinal cord temperature but were depressed again, when local temperature was further decreased. The big phasic α-motoneurons were maximally activated at local temperatures be-

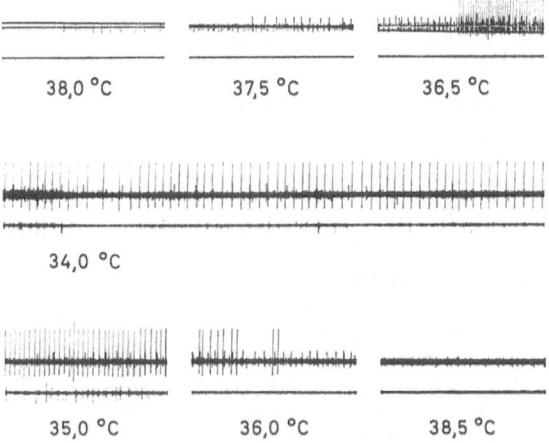

Fig. 16. α-motor (big spikes) and γ-motor (small spikes) unit activity in a L_7 ventral root filament of an anesthetized cat as influenced by local spinal cord temperature. Upper recording trace: ventral root activity, lower recording trace: electromyogram of the hind leg. With decreasing local temperature (upper row) γ-units are activated at slight and α-units at moderate degrees of hypothermia. During cooling (local temperature 34° C, middle row) the α-motor unit is still continuously active, though at reduced frequency, while γ-motor activity is depressed. The electromyogram indicates shivering. During rewarming (lower row) the α-unit activity is first increased but disappears at 36.5° C, when the γ-unit is again active. The control state of activity is reached at a local temperature of 38.5° C. (KLUSSMANN, 1969)

tween 35 and 30° C and tended to respond tonically during spinal hypothermia. As shown by STELTER and KLUSSMANN (1969) the smaller tonic α-motoneurons reached maximum activity at local temperatures above 35° C. Heating the spinal cord above the normal core temperature had, as a rule, a depressing effect on motoneuron activity. As pointed out in a preceding Paragraph (1.2.1.), the cold induced changes of γ-motor activity may reduce the dynamic sensitivity of the muscle spindle receptor and may therefore increase the probability that the proprioceptive control loop becomes unstable. Activation of the α-motoneurons by local cooling has to be regarded as an additional factor which facilitates oscillations in the proprioceptive feedback system. From these and from related studies KLUSSMANN has arrived at the conclusion that local cooling effects on spinal motoneurons contribute to the activation of the shivering mechanism in case of a fall of core temperature below its normal level (KLUSSMANN and PIERAU, 1972).

Investigations of PIERAU et al. (1970a) on the pupillomotor responses during brain cooling in the pigeon have confirmed that basically non-specific cooling effects on the neurons controlling somatomotor activity do in fact facilitate the generation of a tremor. Spontaneous oscillations of the striated pupillary muscles were evoked by brain cooling. Since brain cooling in pigeons does not evoke regular shivering (RAUTENBERG et al., 1972), this tremor activation has to be ascribed to the effects of local hypothermia on the motor efferents.

In summary, the investigations of KLUSSMANN and coworkers have inferred that local thermal influences on spinal somatomotor efferents tend to modulate the efferent thermoregulatory drives governing the spinal shivering mechanism.

Even minute changes of spinal cord temperature may be effective, because they lead to measurable changes of motoneuron excitability as will be shown in a subsequent paragraph (1.3.3.). The question remains how much in quantitative respects the thermosensitivity of the motoneurons may contribute to the control of the shivering mechanism.

Local temperature effects on autonomic spinal efferents do not seem to contribute to the vasomotor responses seen during spinal thermal stimulation. Investigations in acutely spinalized dogs showed that spinal cord cooling resulted in a depression of sympathetic activity, while spinal heating activated the sympathetic efferents (SIMON et al., 1970). These effects are opposite to those observed in cutaneous sympathetic efferents of intact animals. In chronically spinalized dogs and rabbits the cutaneous sympathetic efferents were found to respond with increased discharges to spinal cord cooling (WALTHER et al., 1971a, b). However, this could not be due to a non-specific temperature effect on the preganglionic neurons because the activity of the visceral sympathetic efferents was simultaneously depressed. As pointed out before (Paragraph 1.2.3.) this differential vasomotor response of chronically spinalized animals to spinal cooling probably originated in subsidiary spinal thermoregulatory centers.

1.3.2. Temperature Effects on Ascending Spinal Neurons

In animals with an intact central nervous system, the specific function of spinal thermosensitivity in temperature regulation appears to be mainly achieved by the transmission of temperature signals from the spinal cord to the supraspinal thermoregulatory centers. Thermal panting evoked by spinal cord heating (activation of an effector system which is controlled by the supraspinal respiratory center) gave the first unequivocal support to the idea that thermosensitive ascending spinal neurons might be involved in the generation of thermoregulatory responses (JESSEN, 1967). Transection experiments (WÜNNENBERG and BRÜCK, 1968a; KOSAKA et al., 1969b) indicated that the interruption of the spinal anterolateral tracts abolished or reduced thermoregulatory effector responses to spinal thermal stimulation. The relevant spinal temperature signals were apparently conducted in ascending neurons running in these tracts.

Microelectrode recordings from single units in the rhombencephalic part of the medial lemniscal pathway were carried out by WÜNNENBERG and BRÜCK (1968c, 1970) in guinea pigs, in which the lower cervical and upper thoracic segments of the spinal cord were selectively heated. These investigations presented the first direct evidence for afferent transmission of "warm" signals originating in the spinal cord. As shown by Fig. 17, the neurons responded to spinal heating with a sustained increase of discharge rate which was preceded by a dynamic overshoot during the time of rapid temperature rise. The static temperature response relations observed in 8 single units exhibited varying degrees of nonlinearity. The average response could be satisfactorily described as an exponential rise of discharge rate with spinal canal temperature corresponding to a Q_{10} of about 11 in the temperature range between 38 and 45° C.

In anesthetized cats ascending neurons which responded to temperature changes of the lower cervical and the thoracolumbar spinal cord could be identified by microrecording from single axons of the spinal anterolateral tracts at the upper cervical level rostral to the stimulated sections of the spinal cord (SIMON and IRIKI, 1970, 1971a, b). High spinal transections were performed in these in-

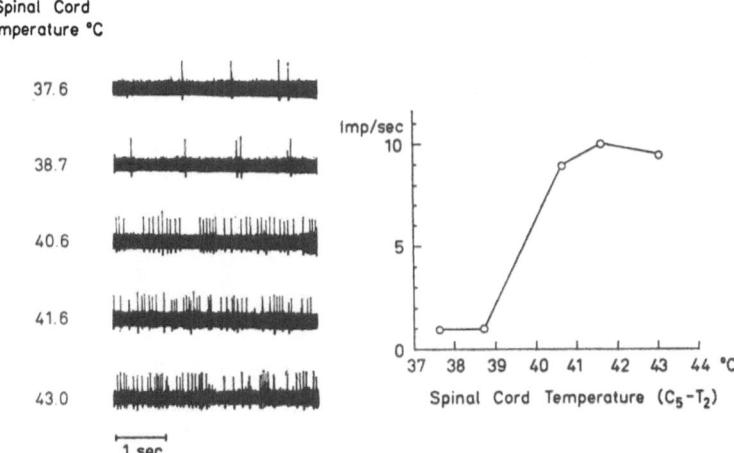

Fig. 17. Warm sensitive ascending neuron recorded from the spinothalamic tract of an anesthetized guinea pig at the medullary level. Warm stimulation of the lower cervical and upper thoracic spinal segments. Left: sections from the original recordings at various states of warming. Right: static discharge rate plotted against spinal cord temperature. (WÜNNENBERG and BRÜCK, 1970)

vestigations in order to prevent recordings from descending neurons which could have been influenced by the supraspinal control centers in response to the spinal thermal stimulus. Two clearly separated sets of thermosensitive neurons could be distinguished according to their static responses to temperature displacements within the spinal canal (SIMON and IRIKI, 1971a, b). Those neurons which responded to spinal heating above normal with a sustained rise of discharge rate were termed "warm sensitive", while the term "cold sensitive" was applied to the neurons being activated by spinal cord cooling below normal core temperature. This classification corresponds to the definition given by BLIGH and JOHNSON (1973), but is regarded merely as descriptive.

About half of both warm and cold sensitive ascending spinal neurons exhibited dynamic components in their responses to temperature changes of the spinal cord. In this respect the discharge characteristics resembled those of cutaneous cold and warm receptors. It is an open question whether the presence or absence of dynamic responses marks functional differences, especially since rapid changes of spinal cord temperature are not likely to occur under natural conditions.

The static responses to spinal temperature displacements of 20 warm sensitive and 20 cold sensitive ascending spinal units were followed up in cats over a range of thermode perfusion temperatures between 14 and 47° C. Temperature measurements within the spinal canal at different distances from the thermode revealed that the changes of neuronal activity were correlated best with the temperature close to the spinal canal thermode (SIMON and IRIKI, 1971b). Therefore, the average responses of the neurons were related to this local temperature which varied with the perfusion temperatures in a range between 22 and 43° C. As shown by Fig. 18, the average temperature-response relationships were found to be non-linear for both groups of neurons in the investigated temperature range. The results suggested that the average temperature signals produced by

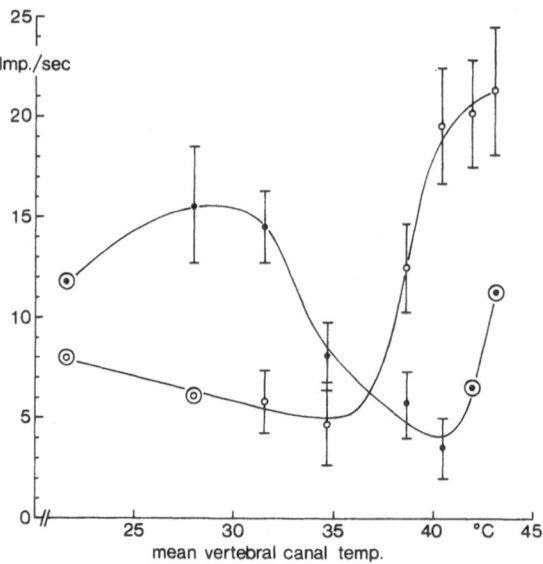

Fig. 18. Average static responses of cold sensitive (dots) and warm sensitive (circles) ascending spinal units recorded in the anterolateral tracts at C_3–C_5 in anesthetized cats with spinal transection at C_1/C_2. Thermal stimulation of the thoracic and lumbosacral spinal cord by means of a water perfused spinal canal thermode. Results from 20 warm sensitive and 20 cold sensitive units. Mean values with standard errors, symbols without standard error marks indicate mean values obtained from less than 6 units. Ordinate: discharge rate, abscissa: average spinal canal temperature measured close to the thermode. (Simon and Iriki, 1971 b)

these two populations of thermosensitive neurons might follow two bellshaped curves with maximum activities between 25 and 30° C for the cold sensitive and above 43° C for the warm sensitive neurons. The steepest and approximately linear parts of both response curves covered the range between 35 and 41° C and intersected approximately at normal core temperature. Average sensitivities in this range amounted to 5.6 Imp./sec/° C in warm sensitive and to −2.4 Imp./sec/° C in cold sensitive units corresponding to Q_{10}-values of 18 and of 0.12 (or "−8.5"). These sensitivities were of the same order of magnitude as those described for hypothalamic heat and cold sensitive neurons (Thauer and Simon, 1972).

The location of the recorded ascending spinal thermosensitive units was assessed by micromarking. Fig. 19 shows that the recording sites of both types of neurons were found in the peripheral parts of the anterolateral tracts. The distribution of cold and warm sensitive units seemed to be not grossly different. This location confirmed that temperature dependent activity had been recorded from afferent axons conducted presumably in a "protopathic" sensory pathway (Simon and Iriki, 1971 b).

The described results by themselves do not allow definite conclusions about specific functions of both populations of thermosensitive neurons to be made. However, according to their location it is reasonable to assume that they are driven by thermosensitive structures mediating basically afferent or sensory

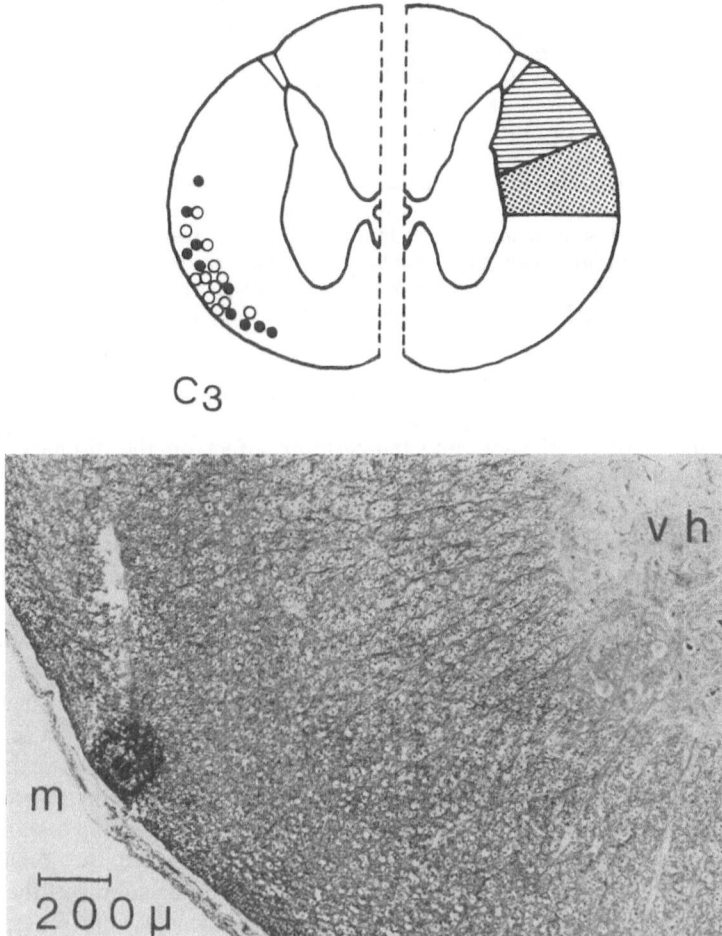

Fig. 19. Recording sites of warm sensitive (circles) and cold sensitive (dots) ascending units in the anterolateral tract as determined by micromarking (left side of the schema; the right side shows the location of the spinocerebellar tracts at the level of C_3 according to OSCARSSON, 1965). The microphotograph demonstrates a transverse section from the 3rd cervical spinal segment with a micromark (m). (SIMON and IRIKI, 1971 b)

informations. They further fulfill basic requirements of a "measuring device" for spinal cord temperature. Thus, they might represent the "input channels" through which spinal thermal stimuli would exert their influences on the supraspinal control centers of the temperature regulation system. However, evidence is at present only circumstantial that central warm and cold sensitivity is in fact established by two different sets of thermosensitive neurons with positive and negative temperature coefficients.

The generally accepted view that the protopathic sensory pathway of the spinal cord carries the thermal input from the skin has raised the question of possible interactions of cutaneous with central temperature signals at the spinal level. In fact, a high degree of convergence between spinal and cutaneous temperature signals has been observed in the ascending spinal cold sensitive neurons of anesthetized spinalized cats (SIMON, 1972). Currently, the modes of interaction at the spinal level appear less multifarious than convergence of temperature signals at the midbrain and hypothalamic levels (HELLON, 1972b). However, interaction at the spinal level was found to be less obvious in anesthetized cats with intact connections between the spinal cord and the brain (unpublished observations). Therefore, conclusions are not yet possible as to what degree the combined effects of spinal and cutaneous thermal stimuli on thermoregulatory effectors might be established by their interaction in the spinal cord. Conclusions on the significance of temperature signal convergence at the spinal level are further limited by the observation that non-thermal afferent inputs were found to influence many of the ascending spinal thermosensitive units (SIMON and IRIKI, 1971b). Further analysis of the modes of signal interaction at the spinal level might, however, help to elucidate the mechanisms by which the signals from different thermoreceptive areas of the body are combined to a common input.

1.3.3. Cellular Mechanisms of Neuronal Thermosensitivity

From a general point of view, any temperature effect on the discharge rate of a neuron must be based on the thermal susceptibility of cellular mechanisms determining the excitability of neural membranes. With regard to the specific character of central nervous thermosensitivity it might well be that it is established by the particular structure of neural networks which otherwise exhibit "non-specific" thermosensitivity as do more or less all excitable structures. Even if specific thermoreceptive structures should exist (e.g. dendritic ramifications generating temperature dependent "receptor potentials"), this would be difficult to assess as long as the thermal influences on pre- and postsynaptic mechanisms of excitation and inhibition remain unknown. As mentioned in a preceding paragraph (1.3.1.), signal transmission in spinal reflex arcs (synaptic function) is temperature dependent. The observations by KLUSSMANN and coworkers of a differential activation of motor efferents by spinal cooling have shown that the motoneurons are by themselves thermosensitive. Therefore, the spinal motoneuron may be chosen as the model subject of investigations aiming at the evaluation of the cellular mechanisms on which spinal or, in a broader sense, central nervous thermosensitivity is based.

The analysis of the properties of lumbar α-motoneurons in cats by means of the intracellular recording technique has confirmed that a silent motoneuron can be activated in situ by lowering its temperature. It was further shown that both mono- and polysynaptic excitatory postsynaptic potentials (EPSPs) which were subliminal at normal core temperature could fire an action potential if local temperature was only slightly decreased (PIERAU et al., 1969a). An experimental condition in which this effect is most obvious is demonstrated in Fig. 20. In this experiment the time distance between the paired stimuli was chosen so that the second EPSP overlapped the declining part of the first EPSP and, thus, had a slightly greater amplitude. Stimulus strength was adjusted so that the second EPSP was just subliminal at normal spinal cord temperature. A slight fall of local temperature by 0.1° C was sufficient to induce occasional firing of the neuron by the second EPSP. With decreasing spinal cord temperature the second EPSP became regularly supraliminal, and eventually even the first EPSP could fire the neuron. This cold induced increase of motoneuron excitability was fully reversible by rewarming.

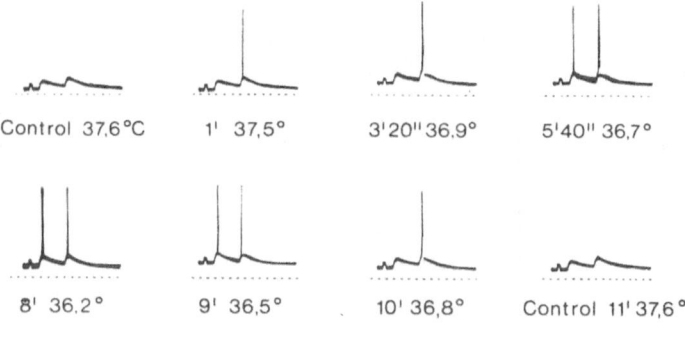

Control 37,6 °C   1' 37,5°   3'20'' 36,9°   5'40'' 36,7°

8' 36,2°   9' 36,5°   10' 36,8°   Control 11' 37,6°

Cal.: 5 mV, 2 msec

Fig. 20. Effects of local spinal cord cooling on intracellularly recorded membrane potential, excitatory postsynaptic potential (EPSP) and action potential of a lumbar motoneuron in the anesthetized cat. Double shock orthodromic stimulation of the biceps semitendinosus nerve. In each recording 8–10 traces are superimposed. Time after the onset of cooling and local intraspinal temperature are indicated. Note the slight decrease of the resting membrane potential with falling local temperature as indicated by the increasing distance between the potential baseline and the stippled reference line. For details see text. (PIERAU, 1971; in: KLUSSMANN and PIERAU, 1972)

Several mechanisms have to be considered by which spinal cooling may cause an increase of excitability. An increase of the resistance of the postsynaptic membrane (evaluated from the potential changes following intracellular current injections) seems to be the most important factor which accounts for the increase of EPSPs during cooling (PIERAU et al., 1969a). The importance of this cold induced increase of membrane resistance for the increase of excitability is documented by the finding that the latency of spike generation after intracellular injections of constant depolarizing current pulses was exponentially related to local spinal cord temperature (Fig. 21). This result has to be expected if a linear increase of membrane resistance with decreasing membrane temperature is the main parameter determining excitability during temperature changes (PIERAU et al., 1971). Further, a slight depolarisation of the motoneuron membrane commonly developed during spinal cord cooling (PIERAU et al., 1969a). This resulted presumably in a reduction of the voltage difference between the resting membrane potential and the threshold potential, since the level of the excitation threshold appeared to be not affected by cooling. Increase of excitability by membrane depolarisation seems, however, to be of secondary importance because it did not correlate as well with the observed changes of excitability as did the increase of membrane resistance (PIERAU, 1971).

As pointed out by PIERAU (1971), thermal influences on membrane permeability for K^+ and Na^+ could sufficiently explain both the changes of membrane resistance and of resting membrane potential during spinal cooling. In addition, effects on electrogenic active transport mechanisms must be considered (SPERELAKIS, 1970), although the observed Q_{10}-values of membrane resistance and resting potential do not call for the participation of energy dependent mechanisms.

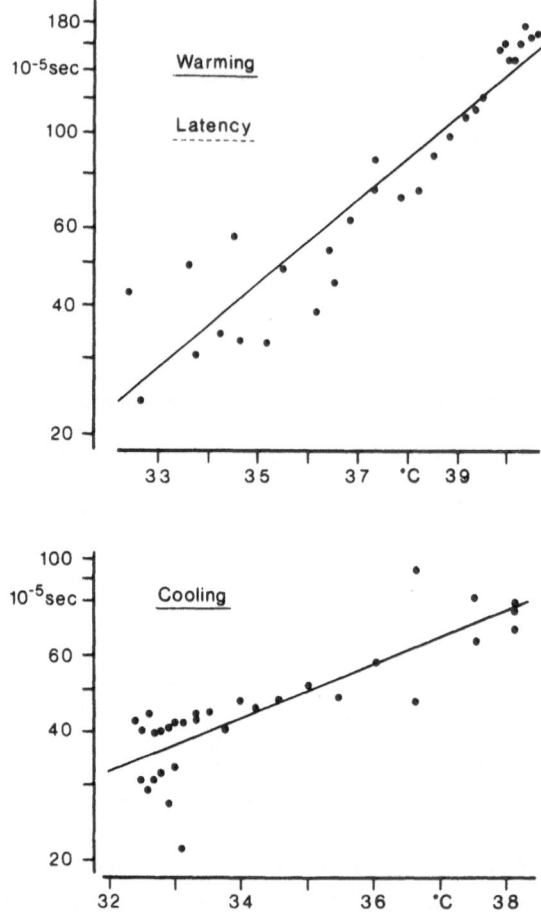

Fig. 21. Latency (ordinate, logarithmic scale) of action potentials of a lumbar motoneuron after injection of a constant depolarizing current in relation to the local intraspinal temperature (abscissa). Lower diagram: values obtained during cooling (falling spinal cord temperature), upper diagram: values obtained during rewarming (rising spinal cord temperature). (Pierau et al., 1971)

SPERELAKIS (1970) has discussed the problem of thermosensitivity of excitable cells under the aspect of how oscillatory mechanisms responsible for spontaneous repetitive discharges may be influenced by temperature. The experimental data about temperature dependent biophysical parameters involved in impulse generation which are compiled in his study show that during cooling of excitable cells of various tissues in various species, membrane resistance is increased and resting potential is decreased or unchanged. Among other factors changes of ionic membrane conductance and thermal effects on electrogenic active transport mechanisms are discussed as the underlying processes. Apparently, the thermal effects on motoneuron activity which were evaluated by PIERAU and coworkers are based on membrane properties which are common to most excitable structures. These properties might also constitute specific thermosensitivity. The tentative explanation of the functions of warm and cold receptors which is given by SPERELAKIS on the basis of the described temperature effects on membrane properties is guided by this view.

Several further observations of PIERAU and coworkers fit in with the concept that membrane resistance and resting membrane potential are the main electrical parameters responsible for thermally induced changes of motoneuron excitability. For instance, increase of resistance and membrane depolarisation during cooling should also increase the inhibitory postsynaptic potentials (IPSPs). This was confirmed by recording from lumbar motoneurons in which IPSPs were evoked by afferent nerve stimulation or by antidromic ventral root stimulation for activation of recurrent inhibition (PIERAU et al., 1969b, 1970b; PIERAU and SPAAN, 1970). PIERAU and SPAAN (1970) could demonstrate that the increase of recurrent IPSPs during cooling was not caused by increased activity of the Renshaw cells and had consequently to be ascribed to the postsynaptic temperature effects. Another parameter which was found to be regularly influenced by local cooling was postexcitatory hyperpolarisation of spinal motoneurons (PIERAU and KLUSS-MANN, 1971). This effect can also be explained by the increased membrane resistance.

The well known increases of amplitude and duration of the action potentials of cooled neuraxons must further be considered as possible sources of increased excitability since they may lead to increased and prolonged transmitter liberation. However, observations of PIERAU (1971) which are in accordance with findings of SUDA et al. (1957) and of KATZ and MILEDI (1965) indicate that this presynaptic effect of cooling probably does not come into play until local temperature has fallen by several degrees centigrade.

Warming of the spinal cord resulted in a depression of activity of α-moto-neurons (KLUSSMANN, 1969). The effects of warming on membrane resistance and resting potential were, in general, opposite to those of cooling (PIERAU et al., 1969a, 1971). Consequently, a reduction in amplitude and duration of both EPSPs and IPSPs were found during spinal hyperthermia (PIERAU, 1971).

From the described results a broad spectrum of possible interactions between the temperature effects on excitatory and inhibitory processes at the postsynaptic membrane may be deduced. The observations of KLUSSMANN (1969) that moto-neurons are excited by spinal cooling, while, on the other hand, discharge rate is subsequently reduced with falling local temperature (see Fig. 16), suggests a tentative hypothesis of how these interactions might account for a "bell-shaped" relation between discharge rate and local temperature in a "cold sensitive" neuron. This hypothesis is schematically illustrated by Fig. 22 according to an experimental condition as described by Fig. 20. At normal temperature (left figure), only the second out of two consecutive EPSPs reaches the threshold of excitation. During moderate hypothermia (middle figure), membrane depolarisa-tion and EPSP enlargement due to increased membrane resistance make the first EPSP surpass the threshold of excitation, i.e. the discharge rate is increased (doubled in this example). At this state of hypothermia, the slight increases of IPSP (shaded area) and of after-hyperpolarisation (hatched area) are not sufficient to suppress the second spike. However, this will occur during stronger cooling (right figure), and, as a consequence, the discharge rate is reduced again (PIERAU and KLUSSMANN, 1971).

The differential activation of motoneurons by spinal cooling which has been described by KLUSSMANN and coworkers further indicates that the local temperature at which maximum activation is reached may differ in individual neurons, possibly in relation to the cell size (KLUSS-MANN, 1969). It is, therefore, conceivable that, depending on the relative "weights" of the

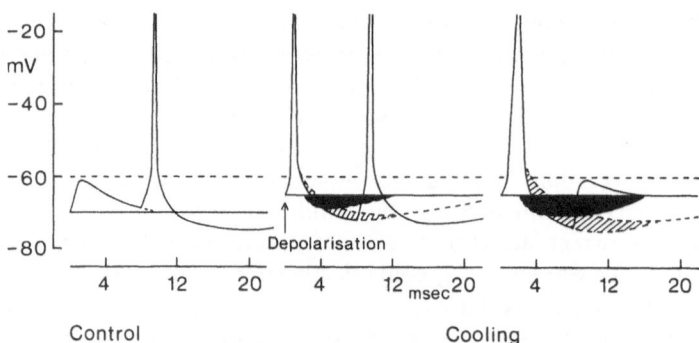

Fig. 22. Schematic survey of the changes of EPSP, IPSP and after-hyperpolarisation occurring in a cold activated spinal (moto-)neuron during spinal cord hypothermia. For details see text. (PIERAU and KLUSSMANN, 1971)

temperature effects on the postsynaptic electrical parameters (EPSP, IPSP, after-hyperpolarisation), maximum excitability might already be attained within or even above the normal range. As an example for spinal neurons with maximum activity above normal core temperature, the Renshaw cell of the cat has been evaluated (PIERAU and SPAAN, 1970). In this way, "warm sensitivity" of a neuron might be established by the same temperature effects which account for the cold sensitivity of the α-motoneurons. This is, however, at present only a speculation which lacks experimental support.

1.4. Synopsis of Thermoregulatory Functions Established at the Spinal Level

The investigations which have been reviewed in this chapter have definitely established the hypothesis that spinal thermosensitivity is specifically related to the temperature regulation system. They have furthermore revealed that temperature changes of the spinal cord may influence the thermoregulatory effector responses in different ways with and without participation of supraspinal control centers. The spinal mechanisms which appear to be involved will be classified in the following paragraphs according to the functions which would be fulfilled by them in a biological control system designed along the principles of a technical controller.

1.4.1. Input Function: The Spinal Cord as a Temperature Sensor

As mentioned previously, the importance of signal transmission from the spinal cord to supraspinal parts of the central nervous system was first indicated by the activation of the panting response during spinal cord heating (JESSEN, 1967). It was further confirmed by the observations that spinal heating failed to suppress shivering in guinea pigs (WÜNNENBERG and BRÜCK, 1968a) or to elicit thermal tachypnea in rabbits (KOSAKA et al., 1969b) if the anterolateral tracts of the spinal cord had been transected rostral to the thermally stimulated sections of the spinal

RABBIT

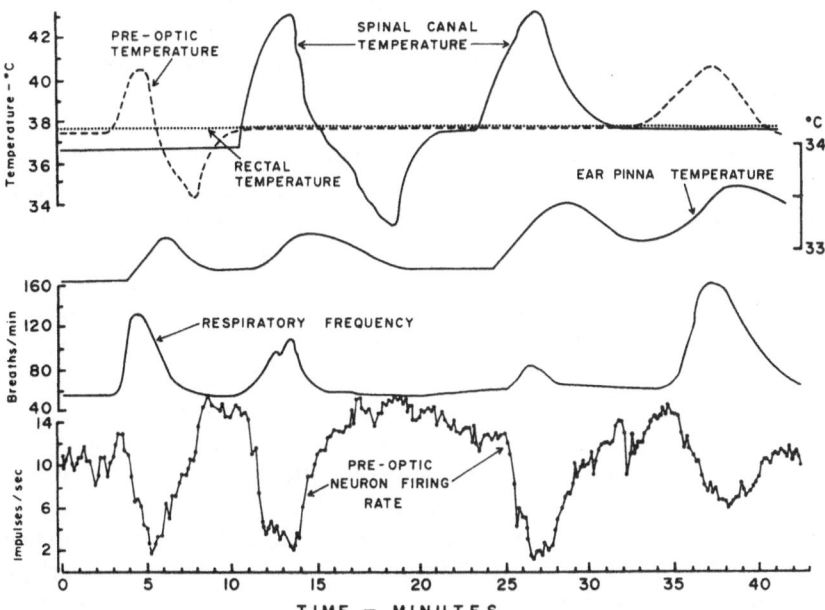

Fig. 23. Firing rate of a preoptic neuron as influenced by local preoptic and spinal cord heating. Note that heating of both areas evokes thermoregulatory responses. (GUIEU and HARDY, 1970b)

cord. On the other hand, decortication did not substantially affect the autonomic thermoregulatory responses (KOSAKA et al., 1969a). These findings inferred that the spinal temperature signals had to be conducted to the brain stem level in order to evoke the full spectrum of responses of the temperature regulation system. WÜNNENBERG and BRÜCK (1970) have arrived at the same conclusion from their observation that the inhibitory effects of spinal cord heating on shivering could be abolished by circumscribed lesions in the anterior hypothalamus by which the input channels ascending from the spinal cord were apparently interrupted.

It has already been pointed out that observations on any type of seemingly specific neuronal thermosensitivity can be related, at the present state of knowledge, only by analogy to the concepts about central nervous thermosensitivity as derived from the analysis of effector responses. With this restriction in mind, the observations on ascending spinal thermosensitive neurons are in accordance with the view (SIMON, 1967) that the input signals contributed by the spinal cord are, in qualitative respects, analogous to those contributed by other central thermoreceptive structures. Single unit recordings in the preoptic hypothalamic region (GUIEU and HARDY, 1970b) and in the caudal hypothalamus (WÜNNEN-BERG and Hardy, 1972) have confirmed that hypothalamic and spinal thermal stimuli may affect the discharge rates of brain stem units in the same way. Fig. 23

demonstrates this equivalence of actions at the preoptic hypothalamic level. Recent neuronal models describing central thermal inputs into the central controller have taken account of this view (Bligh, 1972; Hammel, 1972). Accordingly, hypothalamic, spinal and other central thermal inputs have been handled largely in the same manner in the control equations of Hardy and Guieu (1971) and of Hardy (1972b).

Regarding all hitherto available information, afferent conduction of temperature signals from the spinal cord to the hypothalamus (Thauer, 1968) appears as the most important mechanism by which spinal thermosensitive structures could participate in temperature regulation. Therefore, it is justified to emphasize the function of the *spinal cord as a sensor for deep body temperature* in investigations concerned with the role of the spinal cord in temperature regulation of animals under natural conditions (see Chapter 2). The term "sensor" is used in this context to characterize a circumscribed or anatomically defined region (the skin or the hypothalamus), in which thermoreceptive structures with a specific function are localized (Jessen and Mayer, 1971). Warm and cold sensitivity have been ascribed to the spinal temperature sensor, since both heating above and cooling below normal core temperature are effective in evoking appropriate effector responses. The discovery of two sets of thermosensitive ascending spinal neurons, a cold sensitive and a warm sensitive one, suggests that the informations about spinal cord temperature are provided by two separate though overlapping signals inputs rather than by a continuous signal covering the range from hypo- to hyperthermic temperatures. However, this question is not definitely settled, neither for the hypothalamus nor for the spinal cord.

1.4.2. Controller Function: The Spinal Cord as a Site of Subsidiary Control Functions in Temperature Regulation

As mentioned in the introduction, the idea of extracerebral temperature regulation which can be traced to Goltz (Goltz and Ewald, 1896) may be regarded as the starting point for all the experimental efforts which have eventually led to the evaluation of spinal thermoregulatory functions. This idea had for the first time been incorporated into a concept of temperature regulation by Thauer (1939).

Experimental observations (Dworkin, 1930; Popoff, 1934; Thauer, 1935; Thauer and Peters, 1938), clinical reports, and theoretical considerations concerning the "plasticity" of central nervous functions (Bethe, 1931) had led Thauer (1939) to propose a system of temperature regulation which, though governed by the hypothalamic heat regulation center, was composed of multiple integrating mechanisms widespread in the central nervous system. These integrating mechanisms were supposed to function to a more or less limited degree independently from the hypothalamus, especially if hypothalamic thermoregulatory function was disturbed by gradually developing pathological processes. This hypothesis was able to explain the considerable diversity of results obtained in even the most systematic ablation and destruction studies on the hypothalamus. Since, however, the prevailing experimental efforts at that time were devoted to the establishment of the hypothalamic thermoreceptive and integrative functions in thermoregulation, Thauer's ideas were temporarily bypassed (Hardy, 1961). In retrospect, it appears that quite a few of these experimental efforts, in particular the investigations of Keller and associates (Keller, 1930, 1938; Keller and McClaskey, 1964), have supported rather than opposed Thauer's view.

After the discovery of spinal thermosensitivity (SIMON et al., 1963b) the concept of extrahypothalamic thermoregulation was resumed by THAUER (1964) who suggested with regard to spinal neurons that "synapses concerned with the mediation of thermoregulatory reflexes or formations of cells exerting control functions might exist outside of the hypothalamus." The observations on cold shivering in spinalized (SIMON et al., 1966; KOSAKA et al., 1967; KOSAKA and SIMON, 1968a, b) and midbrain transected animals (SIMON et al., 1968; CHAI and LIN, 1973) have, in principle, confirmed the existence of extrahypothalamic control mechanisms for this important effector response. The observations of RANDALL et al. (1966) on sweating in the spinal man under external heat loads may likewise be regarded as experimental support for the concept of "spinal thermoregulation", although objections might arise from the finding that an increase of skin temperature may directly stimulate the sweat glands (BULLARD et al., 1970). However, investigations on the vasomotor responses of chronically spinalized dogs and rabbits (WALTHER et al., 1971a, b) have further indicated the existence of thermoregulatory control functions at the spinal level. In summary, the reported results support the concept of extracerebral thermoregulation which had been deduced by THAUER (1935) and by THAUER and PETERS (1938) from their observations in rabbits with chronical, high spinal or midbrain transections.

According to our present knowledge, it cannot be decided to what extent the spinal control functions in temperature regulation might contribute in quantitative respects to temperature regulation in *intact* animals. A major theoretical implication of "spinal temperature regulation" may be seen in its phylogenetic aspects (THAUER, 1967). Spinal thermoregulatory control functions may further be taken into consideration as "model mechanisms" (Simon, 1967; THAUER and SIMON, 1972), which are more easily accessible to neurophysiological analysis than the corresponding hypothalamic structures which are interwoven in a network of neurons engaged in a multitude of autonomic control functions.

1.4.3. Output Function: The Spinal Cord as an Output Amplifier

This function may be inferred from the statement of KLUSSMANN and PIERAU (1972) that α-motoneuron excitability is inversely related to spinal cord temperature. As pointed out by these authors, even slight temperature changes within the range of natural core temperature variations will measurably influence motoneuron responsiveness to excitatory stimuli. With respect to temperature regulation, a slight drop of core temperature would not only induce an increased drive for shivering due to an increased cold signal input but would also render the motoneurons more excitable by the local temperature effects. Thus, the drive for shivering transmitted to the motoneurons via the shivering pathway (BIRZIS and HEMINGWAY, 1956, 1957) would be locally reinforced. Conversely, a slight rise of core temperature would not only reduce the central drive for shivering but would also render the motoneurons less excitable to this drive. The question that remains to be answered is to what degree does this amplification or attenuation of the hypothalamic output quantitatively influence the shivering heat production. So far, no observations exist which indicate similar local temperature effects on autonomic spinal efferents mediating thermoregulatory effector responses.

2. The Role of the Spinal Temperature Sensor in Temperature Regulation

In the preceding chapter the modes according to which the spinal cord may participate in the control of thermoregulatory effector activity have been discussed. The observation that subsidiary extrahypothalamic thermoregulatory control functions are exerted by spinal structures in lesioned animals may gain increasing interest with proceeding elucidation of the mechanisms by which the high degree of stability, reliability and plasticity of biological control systems is achieved. The role of the *spinal cord as a temperature sensor* needs to be considered in the first place, when the physiological significance of spinal contributions to temperature regulation is being discussed.

Every attempt to evaluate the physiological significance of central thermosensitivity as a whole or of single central thermoreceptive areas requires a concept of how the temperature regulation system works. In other words, a model is required according to which the functions and interactions of central and peripheral temperature sensors may be described in quantitative terms. Models of temperature regulation have in general been designed according to the principles of non-biological systems. For instance technical thermostatic principles, where constancy is achieved by negative feedback control have been applied. The role which models have played in the elucidation of biological thermoregulatory mechanisms has recently been appreciated by Hardy (1972a) who has also pointed out the still growing importance of models for the interpretation of the rapidly increasing amount of experimental data.

Among the recent models of temperature regulation the block diagram designed by Mitchell et al. (1972) appears to be particularly suitable as an approach to compare the thermoregulatory effects induced by stimulations of different thermal sensors. By proposing dynamic negative feedback control (Mitchell et al., 1970), this model avoids the controversial discussion about the generation of the setpoint or reference signal (at least with regard to short-term thermoregulation) and about the site, where the controlled variable is measured. Both functions have intuitively been ascribed to hypothalamic structures. According to the view that biological control systems are characterized by multiple feedback loops, Brown and Brengelmann (1970) have pointed out that the "set temperature" for effector activation by each temperature sensor may be regarded as comprising the inputs of all the remaining temperature sensors at a given experimental condition. Mitchell's model has taken account of the multiple input concept by proposing that the error signal of the control system corresponds to the degree of imbalance between the total of the cold feedback signals and the total of the warm feedback signals which are produced by the parallel cold and warm inputs from all sensors within the body core and shell (Mitchell et al., 1972).

Although the block diagram proposed by Mitchell et al. (1972) does not cover all the problems which will be discussed in this chapter, it is useful for weighing spinal thermosensitivity against the sensitivity of other temperature sensors. Since the model proposes that cold and warm signals of each thermosensitive region contribute parallel inputs into the central controller, the functions of these sensors may be regarded as equivalent differing only in their quantitative contributions. The modes of interaction should be the same for all combinations of sensor stimulation.

2.1. Comparative Evaluation of the Effectivity of the Spinal Temperature Sensor

The classical way to determine the efficacy of a thermoregulatory feedback loop would be to evaluate its open loop gain as $\Delta T_{core}/\Delta T_{sensor}$ according to the engineering technique of opening the control loop. However, two main difficulties impede this approach. The first one arises from the multiple feedback character of the biological thermostat (BROWN and BRENGELMANN, 1970), and the second one results from the physical thermal imbalances which may be generated by the heating and cooling procedures required for stimulating a temperature sensor.

The first difficulty is a fundamental one. Selective thermal stimulation of one thermal sensor will generate opposing feedback signals of the remaining sensors due to thermoregulatory effector activation which will modify core temperature. These opposing inputs will eventually balance the effect of the original stimulus. As a consequence, the "apparent open loop gain" is much smaller than the true open loop gain: apparent open loop gains for the hypothalamic sensor may be deduced from the core temperature changes which are, with a few exceptions (ANDERSSON et al., 1963a; VON EULER, 1964), in the order of -1 or less. Such low gains can hardly account for the stability of body temperature in many homeotherms as, for instance, dogs (HELLSTRÖM and HAMMEL, 1967), goats (ANDERSSON et al., 1962; ANDERSEN et al., 1962; JESSEN and CLOUGH, 1973a), and monkeys (GALE et al., 1970; ADAIR, 1971). A most instructive example for the balancing effect of opposing feedback signals coming into play during sustained stimulation of one sensor and, correspondingly, for a low apparent open loop gain, has been presented by ADAIR (1971). However, even if it were technically possible to control the inputs of all known temperature sensors, the result would still be more or less elusive, due to the possible existence of further unknown thermal sensors in the temperature regulation system (HAMMEL, 1968; BROWN and BRENGELMANN, 1970).

The second difficulty, a conditional one, arises from the fact that the single thermoreceptive elements, nerve cells or nerve endings, are distributed more or less widely in tissues which are insensitive with regard to temperature perception. If the existence of thermosensitive structures in a circumscribed area is to be inferred from opposing changes of core temperature induced by thermal stimulation of this area, the amount of heat extracted from or supplied to the body by the cooling and heating procedures should be negligible or small compared with the heat accumulation or dissipation evoked by the local stimulus. There is no doubt that this precondition is fulfilled in the investigations concerning thermal stimulation of the preoptic and anterior hypothalamic area in which the thermoreceptive structures seem to be particularly concentrated. In the spinal canal, the situation is less favorable, since it encloses a great amount of tissue probably not engaged in temperature perception. Temperature receptors scattered diffusely throughout the body core, as is tentatively proposed by several authors (MITCHELL et al., 1972; SNELLEN, 1972) would, however, not be recognized at all in this way.

For these reasons, sustained opposing changes of core temperature during a sustained thermal stimulation of a circumscribed region will not give a valid estimate of the gain of the stimulated feedback loop. Such changes will only indicate that thermoreceptive structures have been stimulated which contribute perceptibly to the total signal input of the thermoregulatory control system. However, the efficacy of a feedback loop and, in the last analysis, its open loop gain may be estimated from the relationship between the effector response and the intensity of the stimulus applied to the corresponding temperature sensor. Experimental evaluation of the efficiency of a temperature sensor in this way requires additional assumptions about the modes of action of the central controller in temperature regulation.

HAMMEL and associates were the first to follow this course in order to evaluate in quantitative terms the sensitivity of the hypothalamus from the effector responses evoked by selective temperature displacements (Hammel et al., 1963a, b). Hammel has proposed that homeothermia is established by proportional control with hypothalamic temperature as the controlled variable. According to this concept, the relationship between effector activity and stimulus intensity is described by the equation: $R - R_0 = \alpha_R \cdot (T_{sensor} - T_{set(R)})$; $(R \geq R_0)$, where R is the actual effector response, R_0 is the resting effector activity, α_R is the proportionality constant of the stimulus-response relationship, and $T_{set(R)}$ is the actual threshold temperature for effector activation at the given experimental conditions. The coefficient α_R is proportionally related to the open loop gain of the feedback system (Hammel, 1968) and, thus, gives an estimate of the sensitivity of the stimulated feedback loop. Since no formal reasons exist to handle the various thermal inputs differently in a multiple input system, hypothalamic temperature with its set temperatures for effector activation may be replaced in the control equation by the temperatures and set values of each of the known temperature sensors. Proportionality constants may be determined for individual sensor-effector loops from the stimulus-response relationships, provided that all other inputs are kept constant. This latter condition is, as a rule, fulfilled only during a short time after the start of cooling and heating, since changes of core temperature will modify the feedback signals of the remaining sensors (Hammel, 1965).

2.1.1. Relationship between Stimulus Intensity and Effector Response

The contributions of spinal thermosensitivity to the control of two important thermoregulatory effector systems, the metabolic heat production (or oxygen consumption) and the respiratory evaporative heat loss (or panting rate), have been quantitatively evaluated in dogs, pigeons, pigs, and guinea pigs. Investigations on hypothalamic thermosensitivity carried out with the same methods in the same species or even in the same animal have enabled a *comparison of the spinal and the hypothalamic temperature sensors* on the basis of the stimulus-response relationships.

In experiments on conscious dogs Jessen (1971a) and Jessen and Mayer (1971) analyzed the relationship between heat production and respiratory evaporative heat loss on one hand and of hypothalamic and mean spinal cord temperature on the other by determining linear regression equations according to the method of Hellström and Hammel (1967). The proportionality constants were derived from the regression line slopes, while the corresponding set temperature values (or thresholds of effector activation) were determined from the points of intersection with the levels of resting effector activity. The comparative analysis of spinal and hypothalamic thermosensitivity in dogs was carried out in animals which were equipped with both, spinal and hypothalamic thermodes. The stimulus-response relationships obtained in this way from 4 series of experiments in one dog are shown by the Figs. 24 and 25. These figures which are representative for the results obtained in 4 animals indicate that the sensitivities of the hypothalamic and spinal temperature sensors are in the same order of magnitude. Also warm sensitivity is greater than cold sensitivity in both thermosensitive regions.

At constant, thermoneutral ambient air temperatures ($T_a = 18$ to $24°$ C), the regression equations describing the relation between mean spinal cord temperature and *metabolic heat production* were determined in 4 animals. The slopes of the regression lines varied between -0.3 and -0.7 kcal/kg/hr/$°$ C (-0.35 to -0.81 W/(kg $\cdot$ $°$C)), while the threshold temperatures ranged between 37 and 38$°$ C. Maximum heat production evoked by intensive spinal cord cooling amounted to 3-times the resting value. At warm ambient conditions ($T_a = 30°$ C), the spinal cord had to be cooled down to definitely lower temperatures in order to surpass the threshold of shivering. Results obtained in the same animals during hypothalamic cooling with a set of multiple thermodes in the preoptic and anterior hypothalamic region provided a reliable

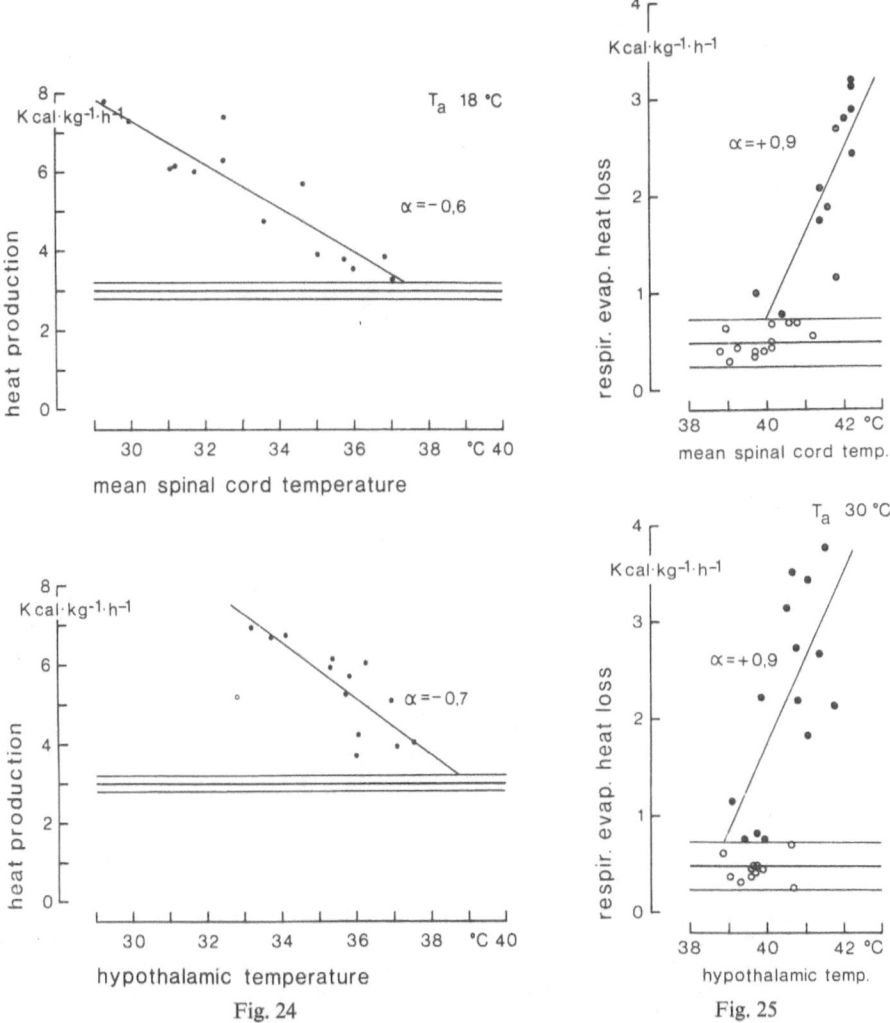

Fig. 24. Stimulus-response relationships between metabolic heat production and the temperatures of the spinal and hypothalamic temperature sensors of a conscious dog at an ambient air temperature of 18° C. (JESSEN, 1971a)

Fig. 25. Stimulus-response relationships between respiratory evaporative heat loss and the temperatures of the spinal and hypothalamic temperature sensors of a conscious dog (same animal as in Fig. 24) at an ambient air temperature of 30° C. (JESSEN, 1971a)

basis for the comparison of spinal with hypothalamic thermosensitivity. The hypothalamic set temperatures for the activation of heat production at thermoneutral ambient conditions (T_a = 18 to 24° C) were found to range between 36.5° C and 38.8° C in three investigated animals. The highest value found for the slopes of the regression lines was −0.7 kcal/kg/hr/°C (−0.81 W/ (kg · °C)). Thus, the thresholds and the proportionality constants of the hypothalamic and the spinal temperature sensors did not differ with regard to metabolic heat production. The relationship between *respiratory evaporative heat loss* and mean spinal cord temperature at

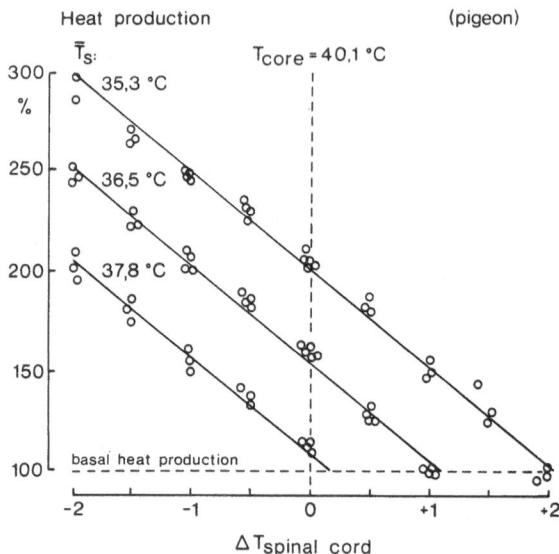

Fig. 26. Stimulus-response relationship between metabolic heat production and the temperature of the spinal temperature sensor of 8 conscious pigeons at 3 different mean skin temperatures. (Rautenberg, 1971)

thermoneutral ambient conditions (T_a = 18 to 24° C) was evaluated in 4 animals. Threshold temperatures between 40° C and 41.5° C were determined from the regression lines, while the slopes were calculated as 1.2 kcal/kg/hr/°C (1.4 W/(kg · °C)). Under warm external conditions (T_a = 30° C), the thresholds were shifted to mean spinal cord temperatures below 40° C, while slopes of 0.9 kcal/kg/hr/°C (1.05 W/(kg · °C)) were calculated. The relation between hypothalamic temperature and respiratory evaporative heat loss at warm ambient temperature was determined in one animal: the slope of the regression line was 0.9 kcal/kg/hr/°C (1.05 W/(kg · °C)), and the threshold temperature was 39 °C. At thermoneutral ambient conditions, the thresholds for activation of respiratory evaporative heat loss by hypothalamic heating were found at temperatures of 41 and 42.5° C in 2 animals. In the narrow range above these high threshold temperatures, heat loss increased steeply with temperature and reliable slopes could not be calculated.

The slopes of the regression lines describing the relation between heat production and evaporative heat loss on the one hand and hypothalamic and spinal cord temperature on the other hand were found to be about equal in Jessen's investigations in dogs. Somewhat greater slopes were evaluated for the hypothalamus in the same species by Hellström and Hammel (1967). Differences in the experimental set-up which might have accounted for this discrepancy between corresponding proportionality constants have been discussed by Jessen and Mayer (1971).

Carlisle and Ingram (1973a) have evaluated the relationship between oxygen consumption and spinal and hypothalamic cold stimuli of varying intensities in the pig. It was found for both central sensors that the metabolic rate was linearly related to cooling intensity. The slope of the stimulus response curve for the spinal temperature sensor appeared to be steeper rather than equal to that for the hypothalamic temperature sensor.

Rautenberg (1971) has determined the relation between metabolic heat production and spinal cord temperature in the pigeon for temperature displacements ranging from $-2°$ C below to $+2°$ C above the normal core temperature of 40.1° C. As shown by Fig. 26, heat production was found to be linearly related to spinal cord temperature in the investigated range. Changes of mean skin temperature changed the threshold values for activation of heat production but did not change the slope. The proportionality constant for shivering as determined from the measurements of RAUTENBERG (1969, 1971) amounted to -1.75 kcal/kg/hr/° C (2.05 W/(kg · °C)). This is a value considerably higher than that found for the hypothalamic and spinal temperature sensors of the dog, which may relate to the fact that the hypothalamus of the pigeon does not contribute inputs to the control of shivering (cf. Fig. 7).

Similar results as in pigeons may be expected in guinea pigs, in which hypothalamic temperature likewise does not seem to influence shivering thermogenesis. The proportionality factor for shivering heat production as influenced by spinal cord temperature has not yet been evaluated in this species. However, from the hyperbolas describing the combinations of skin temperature and spinal cord temperature at different levels of electromyographical shivering (BRÜCK and WÜNNENBERG, 1970) an approximative estimation of the stimulus-response relations between heat production and spinal cord temperature may be obtained by converting shivering activity to metabolic heat production according to the diagram of BRÜCK and WÜNNENBERG (1966). The stimulus-response curves constructed in this way at different subcutaneous temperatures would be not rectilinear, resembling those evaluated by SCHWENNICKE (1972) for non-shivering thermogenesis and hypothalamic temperature. The slopes of the response curves describing shivering heat production would correspond to -1 to -2 kcal/kg/hr/°C at normal core temperature and would, thus, be comparable to those found in the pigeon.

Summing up the investigations, in which the effects of spinal and hypothalamic temperature displacements on metabolic heat production and respiratory evaporative heat loss were quantitatively determined under comparable conditions, it may be stated that their results fit in with the concept that the spinal cord operates as a central temperature sensor comparable to the hypothalamus. The comparison of spinal with hypothalamic thermosensitivity in dogs and pigs suggests that both sensors are about equally effective in these species. This statement should not be taken too strictly, since shortcomings of the methods used for thermal stimulation and for measuring stimulus intensity might easily change the weights in favour of one or the other sensor. It should be further kept in mind that the validity of this statement depends on the validity of the underlying model of proportional control according to which this comparison has been carried out. In particular, rate control components in the effector responses (HARDY, 1965) have not been definitely ruled out. However, even with this restriction in mind it may be said that heat and cold sensitivities of the spinal cord and hypothalamus are of the same order of magnitude in dogs and pigs.—In the pigeon, warm and cold sensitivity of the spinal cord as determined from the effects on metabolic heat production seemed to be predominant in comparison to hypothalamic thermosensitivity. The same conclusion may be drawn with respect to the shivering response of the guinea pig.

Analysis of the effector responses to *simultaneous thermal stimulation of the hypothalamus and the spinal cord* has supported the view of MITCHELL et al. (1972) that these central temperature sensors contribute parallel cold and warm signals to the total input of the central controller. With regard to metabolic heat production spinal cord and hypothalamus were found to interact with each other according to the concept of signal addition. As shown by the left hand diagram of

dog F   air temperature 18 °C

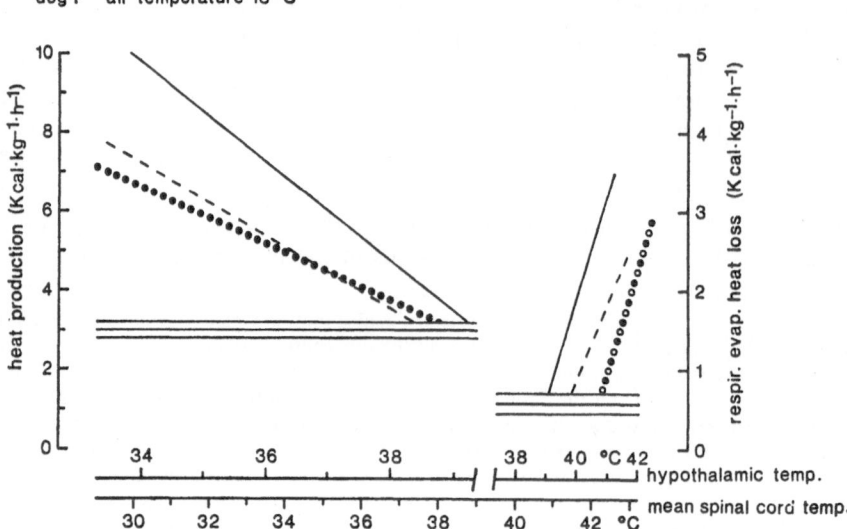

Fig. 27. Stimulus-response relationships between metabolic heat production (left side of figure), respiratory evaporative heat loss (right side of figure) and the temperatures of central temperature sensors of a conscious dog (same animal as in Figs. 24 and 25) at an ambient air temperature of 18° C. Thermal stimulation of the spinal cord: broken line, of the hypothalamus: dotted line, of both sensors: solid line. Since at a given perfusion temperature different hypothalamic and mean spinal cord temperatures were measured, both temperatures are indicated in the abscissa. The common abscissa would be the temperature of the thermode perfusion. (Jessen and Ludwig, 1971)

Fig. 27, simultaneous cold stimulation of both sensors in the dog lowered the threshold for activation of metabolic heat production and increased the slope of the response curve. In accordance with this finding, a constant displacement of the temperature at one sensor induced an approximately parallel shift of the stimulus-response curve of the other sensor (Jessen, 1971 b; Jessen and Ludwig, 1971). However, a simple addition of signals could not be concluded from the analysis of heat loss activation by simultaneous stimulations of both sensors. As shown by the right hand diagram of Fig. 27, simulatneous heating of the hypothalamus and the spinal cord also lowered the threshold for effector activation in comparison to the stimulation of a single sensor. However, the slope of the response curve did not change significantly.

Further observations of Jessen confirmed that metabolic heat production and respiratory evaporative heat loss were differently affected by simultaneous thermal stimulations of the hypothalamus and the spinal cord, especially when opposing stimuli were applied (Jessen and Simon, 1971). Concerning heat production, the results fitted in with the concept that the inhibiting input of the heated sensor was added, in a sense, to the activating input of the cooled sensor. The regression lines describing the relations between heat production and cooling intensity of each sensor were mutually shifted to lower temperatures, when the other sensor was heated to a constant degree. However, cooling of one central sensor did not result in a significant shifting of the regression line describing the relation between respiratory heat loss and the temperature of the other central sensor. This corresponds to the observation that thermal panting induced by intensive heating of the spinal cord or hypothalamus could not be inhibited by even intensive cooling of the remaining central sensor (cf. Fig. 12).

The results of the investigations in dogs on the modes of interaction of spinal and hypothalamic thermal stimuli in determining effector activity may be condensed into the statement that no substantial differences seem to exist between the hypothalamic and the spinal temperature sensors in qualitative as well as in quantitative respects. However, warm and cold signals from both sensors appear to act differently on the mechanisms of heat production and respiratory evaporative heat loss. While the heat accumulating mechanism is about equally susceptible to the activating cold and the inhibiting warm signal input, the heat dissipating mechanism seems to be predominantly under the control of the warm signal input. It is not yet clear, whether this differential susceptibility of heat production and heat dissipation to central cold and warm signals is found only in particular species, such as the dog, or whether it reflects a general characteristic of the temperature regulation system.

The mutual influences on the stimulus-response relationships of central sensors induced by simultaneous, equidirectional or opposing thermal stimulations have been systematically evaluated only in dogs. However, several further reports exist which allow conclusions on the modes of interaction between the hypothalamic and the spinal temperature sensor in other species. With regard to *thermal panting*, the observations of GUIEU and HARDY (1970a) in rabbits suggest a similar preponderance of central warm stimuli as in dogs. On the other hand, the findings of INGRAM and LEGGE (1972a) in pigs do not indicate a preponderance of central warm signals over central cold signals in the control of panting. In the guinea pig, a quantitative analysis of thermal tachypnea indicated an additive interaction of spinal and hypothalamic temperature sensors which were about equally effective (HERRMANN, 1972), but the effects of opposing central thermal stimulations of the hypothalamus and the spinal cord were not tested. As far as the control of *metabolic heat production* is concerned, the investigations of CARLISLE and INGRAM (1973a) in pigs indicate that hypothalamic and spinal temperature sensors interact with each other in this species in a similar manner as in dogs. In guinea pigs (BRÜCK and SCHWENNICKE, 1971) and in pigeons (RAUTENBERG et al., 1972) spinal and hypothalamic temperature sensors do not seem to interact in the control of metabolic heat production. Shivering was predominantly controlled by the spinal cord in both species, while non-shivering heat production in guinea pigs was predominantly controlled by the hypothalamus.

The series of investigations carried out by RAUTENBERG (1969, 1971) in pigeons have not only revealed a linear relationship between spinal cord temperature and heat production, but have further shown, as visualized by Fig. 26, that the *interaction between spinal and peripheral thermal stimuli* occurs along the lines described by HELLSTRÖM and HAMMEL (1967) for the interaction between hypothalamic and cutaneous thermal stimuli. The spinal set temperature for metabolic heat production was shifted to higher values by a rise of mean skin temperature while the proportionality constant did not change. As mentioned already, the observations of JESSEN and MAYER (1971) in dogs concerning the interaction of peripheral with spinal and hypothalamic thermal stimuli likewise seem to follow this rule. CARLISLE and INGRAM (1973a) found the same kind of interaction between spinal and cutaneous thermal stimulation in the control of metabolic heat production in the pig. The way in which their results are presented illustrates the previously mentioned view of BROWN and BRENGELMANN (1970) that each thermal input

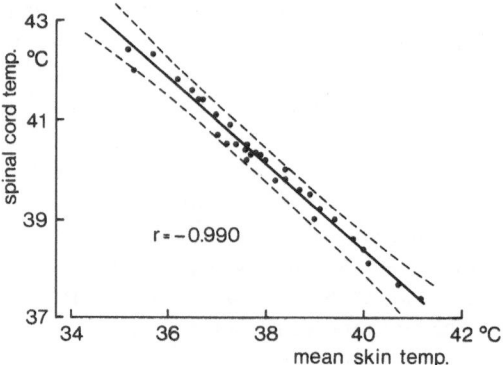

Fig. 28. Combinations of mean skin temperature and spinal cord temperature at the threshold of shivering. Data from 15 conscious pigeons. (Rautenberg, 1971)

may be described equally well as the controlled variable or as a thermal factor changing the setpoint. For example changes of spinal cord temperature in the pig appropriately shifted the "cutaneous set temperature" for the activation of extra heat production.

The interactions between cutaneous and spinal thermal stimuli in pigeons, dogs and pigs may be described as an addition of signals, as it has been proposed by Hammel (1968) for the combined influences of peripheral and hypothalamic thermal stimuli on metabolic heat production and respiratory evaporative heat loss. It is important, however, to emphasize that the interactions of cutaneous and central thermal stimuli do not in any case follow the rule of signal addition. While the parallel shifts of the regression lines of Fig. 26 and the temperature combinations at the threshold of shivering in Fig. 28 indicate a purely additive interaction of spinal and skin temperature signals, the interaction of the skin and spinal cord in the control of shivering in the guinea pig can be described best by a multiplication of the sensor temperatures (Brück and Wünnenberg, 1967a). In this species, the same multiplicative interaction was found in the control of non-shivering thermogenesis by the temperatures of the hypothalamus and the skin (Brück and Schwennicke, 1971). Multiplicative control of effector activity by peripheral and central thermal inputs was most obvious at the thresholds of activation as shown by Fig. 29. This figure further demonstrates the important finding that the roles of spinal and hypothalamic temperature sensors are completely analogous in the control of shivering and, respectively, in the control of non-shivering thermogenesis.

Regarding the discussion of whether the interaction between two different temperature sensors is best described by addition or by multiplication of the signal inputs, the comparison between the Figs. 28 and 29 illustrates the difficulties to give a general answer to this question at the present state of knowledge. Brown and Brengelmann (1970) have pointed out the provisional meaning of mathematical descriptions of input-output relations based on experimental curve fitting. Examples have been reported in the literature which illustrate the limitations impeding the attempts to attain well-established transfer functions for the thermoregulatory control system from the experimental data. For instance, the way in which the data are processed

Temperature combinations at the thresholds of

non - shivering thermogenesis   shivering   (guinea pig)

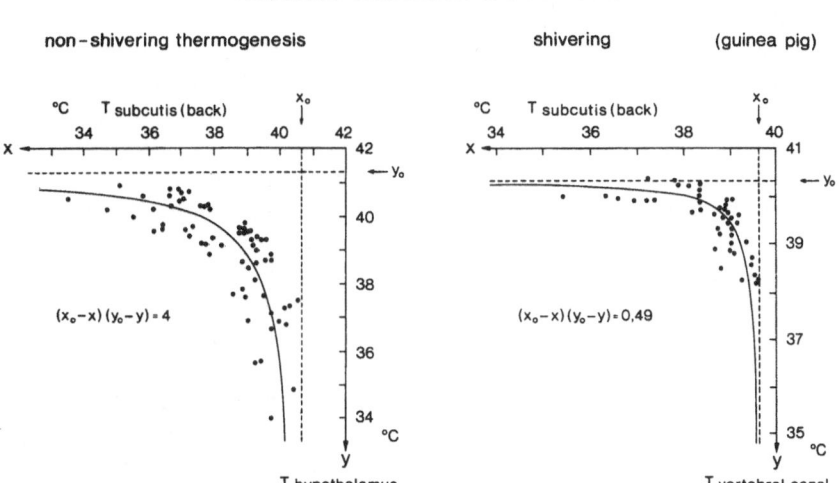

Fig. 29. Combinations of subcutaneous and hypothalamic temperature at the threshold of non-shivering thermogenesis in conscious guinea pigs (left side of figure). — Combinations of subcutaneous and spinal canal temperature at the threshold of shivering in conscious guinea pigs (right side of figure). (BRÜCK and WÜNNENBERG, 1967a; BRÜCK and SCHWENNICKE, 1971)

will influence the conclusions drawn on the underlying coupling function. As an example, depending on the way of evaluation, the data from which HELLSTRÖM and HAMMEL (1967) have inferred an additive interaction of peripheral and hypothalamic thermal stimuli in the control of heat production could also be interpreted as indicating multiplicative control (JACOB-SON and SQUIRES, 1970). Further, mathematical statements about additive or multiplicative interaction are jeopardized by the fact that the discharges of the thermosensitive neurons in the skin and core and, thence, their input signals are in all probability not linearly related to local temperature. A non-linear relationship might also exist between the output signals of the controller and the activity of the thermoregulatory effectors (HAMMEL, 1972). It is, there-fore, not surprising that quantitative studies concerning the interactions of peripheral and central thermal stimuli in the control of sub-systems of the same thermoregulatory effector mechanism may lead to equivocal results. Compare for instance, the response of respiratory evaporative heat loss (HELLSTRÖM and HAMMEL, 1967) and of salivation (HAMMEL and SHARP, 1971) in the dog. SHARP and HAMMEL (1972) have pointed to the non-linearities of the input and output coupling mechanisms of the biological control system as possible causes for a seem-ingly multiplicative interaction of the sensor temperatures.

As a consequence of the reported observations made by BRÜCK and by others, it must be taken into account that the thresholds of effector activation $(T_{set(R)})$ and the proportionality factors (α_R) describing the stimulus-response relations for one sensor according to HAMMEL (1968) may both depend on the thermal states of the other thermosensitive regions of the body. This refers possibly not only to the interaction of central with peripheral temperature sensors but also to the interaction between the diverse central sensors. The model of temperature regulation proposed by STOLWIJK (1970) has assumed an additive and a multi-plicative component in the interaction between temperature sensors. The degree to which each component contributes to the actual stimulus-response relationship, may, however, differ widely, depending, for example, on the species and on the investigated effector.

It is obvious from the contents of this paragraph that the *evaluation of spinal thermosensitivity from the stimulus-response relations* has been faced with several problems concerning the interpretation of the experimental data. It has, however, likewise become apparent that these are largely the same problems which have arisen in corresponding analyses of hypothalamic thermosensitivity. Lacking knowledge about the coupling functions for various control loops and the diversity of modes of interaction between central and peripheral sensors in different species appear as the main factors which impede the attempts to come to a clear-cut statement about the role of the spinal temperature sensor in the thermoregulatory control system. Despite these difficulties it may be stated that *the spinal cord and the hypothalamus have appeared* in the reported investigations *as central temperature sensors with analogous functions*, if the experimental conditions allowed to compare their functions. In quantitative respects, the spinal temperature sensor was found to be about equally effective, less effective or more effective than the hypothalamus in driving heat production and evaporative heat loss, depending on the species and on the investigated effector mechanism. Experimental support has further been contributed to the view of BENZINGER (1969) that the central temperature sensors are, in general, less responsive to cold stimulation than to warm stimulation.

2.1.2. Changes of Core Temperature Induced by Spinal Thermal Stimulation

The stimulus-response relations for metabolic heat production and respiratory evaporative heat loss have elucidated the efficiency of the spinal in comparison to the hypothalamic temperature sensor with regard to the two most powerful effector mechanisms in autonomic temperature regulation. The importance of the spinal temperature sensor for the control of the entirety of the thermoregulatory effectors could be comprehensively evaluated, if it were possible to determine the true open loop gain of the spinal feedback loop from the opposing changes of core temperature induced by selective spinal thermal stimulation. The reasons have been considered why this is not possible. Nevertheless, observations about the apparent open loop gain have contributed important informations about particular properties of the spinal as well as of the hypothalamic temperature sensor.

As demonstrated in Fig. 30 by observations of JESSEN et al. (1972) in the ox at cool ambient conditions, sustained upward displacements of spinal cord temperature resulted in downward displacements of extraspinal core temperature. The fall of core temperature was found to be related to the degree of spinal heat stimulation in an exponential-like manner. At moderate degrees of spinal heating an "apparent open loop gain" of less than -1 indicated the balancing influence of the cold signal feedback from elsewhere in the body core. However, when a certain degree of spinal hyperthermia was transgressed, the spinal warm signal seemed to be no longer balanced: the animals showed a sustained activation of heat loss mechanisms and a progressive decline of core temperature which

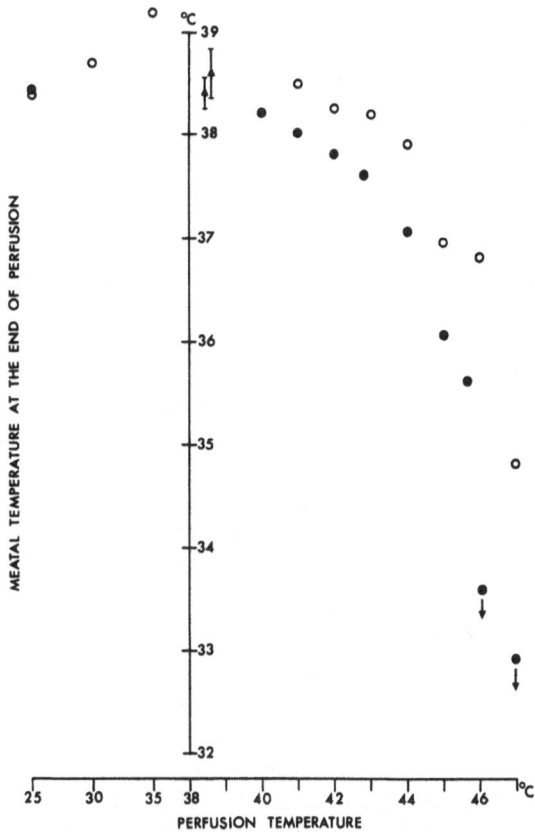

Fig. 30. Core temperatures (meatal temperatures) of two conscious oxen of 234 kg (filled symbols) and 212 kg (open symbols) body weight at the end of thermal stimulation of the spinal cord in relation to the temperature of thermode perfusion. The triangles refer to mean pre-stimulation values (with standard deviations). Arrowed points represent experiments in which meatal temperature failed to stabilize during stimulation time (6 hrs). (JESSEN et al., 1972)

did not show signs of levelling off within the observation time of 6 hrs. As a result, core temperatures as low as 33° C were reached during this time.

This effect of intensive spinal cord heating on extraspinal core temperature in the ox is closely similar to the effect of intensive hypothalamic heating in goats as described by ANDERSSON et al. (1963a). By elevating anterior hypothalamic temperature above 42° C at external cold conditions, extrahypothalamic core temperature could be decreased to values below 30° C. This indicates that the hypothalamic heat signals were not balanced by extrahypothalamic cold feedback signals.

If one tries to describe these findings according to the terminology of MITCHELL et al. (1970, 1972), it may be suggested that the increments of the spinal or hypothalamic rise feedback signals with rising local temperature must have become progressively greater, while the increments of the opposing fall feedback signals with falling core temperature have become progressively smaller. As a consequence, a balance of the rise and the fall feedback signals at tolerable degrees of general hypothermia could no longer be achieved if the local central heat stimulus had surpassed a critical level of stimulus intensity.

As mentioned in Paragraph 2.1.1., a similar disproportion of central heat and cold sensitivity with regard to the control of respiratory evaporative heat loss has been observed in dogs (JESSEN and SIMON, 1971): cooling of one central sensor had no influence on panting induced by heating the other central sensor. However, central cold sensitivity is obviously well developed in the spinal cord and the hypothalamus of dogs, since heat accumulating mechanisms are readily activated by selective or combined cooling of these sensors. Therefore, it may be suggested as an alternative to the just mentioned explanation that the heat loss mechanisms might preferably be controlled by the central warm signals and only little by the central cold signals. Since hyperthermia is a more severe impact on the vital functions, this preponderance of the central warm signals would appear useful.

It is difficult to say, whether VON EULER's (1964) observations in rabbits might be explained in this way. The profound hypothermia induced by even slight increases of hypothalamic temperature seems to indicate a hitherto unique preponderance of central warm signals over central cold signals in this species. Recent investigations on hypothalamic thermosensitivity in rabbits suggest smaller hypothalamic apparent open loop gains (GUIEU and HARDY, 1970a; KLUGER et al., 1973). However, these investigations were not specially concerned with the evaluation of core temperature changes in response to sustained hypothalamic heat stimulation.

An explanation of the findings in oxen and goats according to MITCHELL's model would suggest that central cold sensitivity may be little developed in these species. Actually, JESSEN et al. (1972) could show in the ox that opposing changes of extraspinal core temperature were induced by spinal cooling only at slight degrees of spinal hypothermia (cf. Fig. 30). At stronger spinal cord cooling, the induced spinal fall feedback signal was apparently too small to even balance the amount of heat extracted from the body by the spinal canal thermodes. In the goat similar observations have been made in experiments in which moderate sustained upward and downward displacements of spinal cord temperature were performed (JESSEN and CLOUGH, 1973b). In contrast to spinal heating, spinal cord cooling had only little influence on core temperature at thermoneutral or warm ambient conditions. A more pronounced increase of core temperature was induced by spinal cooling at cold ambient conditions.

Corresponding investigations concerned with the cold sensitivity of the hypothalamus in goats and oxen have been carried out only in the goat (JESSEN and CLOUGH, 1973a). They have revealed a distinct difference between the hypothalamic and the spinal temperature sensors with regard to cold sensitivity. Hypothalamic cold stimuli were as effective as hypothalamic warm stimuli in driving core temperature into opposite directions, at least in the investigated range of moderate heating and cooling. A well-developed responsiveness to hypothalamic cold stimulation in the goat had already been indicated by the observations of ANDERSEN et al. (1962) according to which general hyperthermia developed during prolonged hypothalamic cooling. It remains to be ruled out that the stronger effect of hypothalamic cooling on the heat accumulating mechanisms was not due alone to stimulation of central thermoreceptors but, in addition, also to cooling effects on hypothalamic neurohumoral mechanisms.

If it is taken for granted that in goats and possibly also in oxen spinal cold sensitivity is less developed than hypothalamic cold sensitivity, it is necessary to emphasize that this finding should not be generalized. In pigeons (RAUTEN-BERG, 1969; RAUTENBERG et al., 1972), rats (LIN et al., 1972), and monkeys (CHAI and LIN, 1972), a rise of extraspinal core temperature or even a substantial central hyperthermia could be induced by sustained spinal cord cooling (cf. Fig. 15). In the pigeon the relationship between spinal and hypothalamic cold sensitivity was found to be even reversed in comparison to the goat (RAUTEN-BERG et al., 1972). Brain cooling in pigeons was neither able to increase metabolic heat production nor to induce a rise of extracerebral core temperature. It might be suggested from the observations in goats and oxen on the one hand and in pigeons, rats and monkeys on the other that the degree of spinal cold sensitivity might be inversely related to body size. The observations of RAUTENBERG et al. (1972) further raise the question whether the distribution of central nervous thermosensitive structures in the avian class might have followed a line of evolution different from that in mammals.

2.2. Evidence for Continuous Signalling of Spinal Cord Temperature in the Range of Physiological Core Temperature Variations

Evaluation of the physiological significance of any central temperature sensor from the effector responses to selective thermal stimulation is faced with the problem that thresholds of effector activation exist. As a rule, "dead zones" are found under experimental conditions between the thresholds for the heat accumulating and the heat dissipating mechanisms which cover just that range within which core temperature is normally kept constant. It is difficult to present evidence that the temperature regulation system is continuously active within this range. It is generally supposed that the gap between the thresholds of the two most powerful—and most easily measurable—effector responses of heat production and evaporative heat loss is bridged by effector mechanisms controlling conductive heat loss (skin blood flow, piloerection and postural behavior). However, the continuous, *controlled* function of these effectors is difficult to establish. Accordingly, the evaluation of the continuous function of a central temperature sensor in controlling the function of thermoregulatory effectors in the range of natural core temperature variations is a problem of general interest.

With respect to the spinal temperature sensor this question has been asked with special emphasis, because the thermal stimuli which had been applied in the first exploratory studies on spinal thermosensitivity were clearly outside of the range of physiological core temperatures. Strong spinal heating and cooling had been performed in order to evoke distinct reactions, especially in anesthetized animals. This seemed justified because most of these investigations aimed at the analysis of the effector responses rather than on the evaluation of a possible physiological function of spinal thermosensitivity. However, the impression was gained from these experiments that only gross temperature displacements of the spinal cord were able to elicit thermoregulatory responses. As a consequence, the assumption was suggested that spinal thermosensitivity might constitute a kind of "broad band control" or a "second line of defence" against hypo- and hyperthermia in contrast to the finer "narrow band control" exerted by the hypothalamic temperature detectors (BLIGH, 1966).

2.2.1. Signals of the Ascending Spinal Thermosensitive Neurons

If the thermosensitive neurons ascending from the spinal cord are regarded as a temperature measuring device, their stimulus-response curves support the view that the spinal temperature sensors provide a continuous signal covering the range from hyperthermic to hypothermic core temperature values. The observations of WÜNNENBERG and BRÜCK (1970) suggest that spinal temperature signals would be generated in the guinea pig not only in the temperature range above normal core temperature but also at slight degrees of central hypothermia since the warm sensitive neurons described by them had not reached zero activity at normal core temperature. According to our own observations (SIMON and IRIKI, 1971b), the two sets of cold and warm sensitive neurons found in the cat would be able to measure core temperature continuously in a range extending from considerable degrees of hypothermia to distinct hyperthermia with maximum sensitivity being attained in the range of normal core temperature variations (Fig. 18). GUIEU and HARDY (1970b) have concluded from the effects of spinal thermal stimuli on anterior hypothalamic neurons that the "setpoints" of the spinal thermosensitive structures are found in the range of normal core temperatures as is observed for the "setpoints" of hypothalamic thermosensitive neurons.

2.2.2. Thresholds of Effector Activation

Experiments in which the spinal threshold temperatures for activation of metabolic heat production, cutaneous vasoconstriction and dilatation, and for panting were determined have helped to correct, at least partly, the impression that spinal thermosensitivity can only be concerned with "broad band control" in thermoregulation. Under appropriate external conditions, the spinal threshold temperatures for effector activation were sometimes found within the range of natural core temperature variations (THAUER and SIMON, 1972). In particular, two series of investigations concerned with the interaction of spinal and peripheral thermal inputs in determining the threshold for shivering have indicated a continuous function of the spinal temperature sensor from hypo- to hyperthermic core temperatures. In the pigeon, evaluation of the shivering threshold by electromyography at different combinations of skin and spinal cord temperature resulted in a well defined set of temperature combinations (cf. Fig. 28) in which the spinal cord temperatures were continuously and linearly distributed between 38 and 42° C, i.e. 2° C above and below normal core temperature of the pigeon (RAUTENBERG, 1971). Evaluation of the shivering threshold in the guinea pig in a similar way resulted in a set of temperature combinations (cf. Fig. 29) in which spinal cord temperatures were continuously distributed between approximately 38 and 40° C, i.e. 1° C above and below normal core temperature (BRÜCK and WÜNNENBERG, 1967a). These observations correspond to the result gained from the analysis of the ascending spinal thermosensitive units and support the view that the spinal sensor temperature sensor is continuously operating in the range of natural core temperature variations. In the case of the guinea pig, the conclusiveness of this interpretation might perhaps be questioned because of the non-linear distribution of the pairs of threshold temperatures. However, the linear and very narrow distribution of the threshold temperatures in the pigeon leaves little doubt.

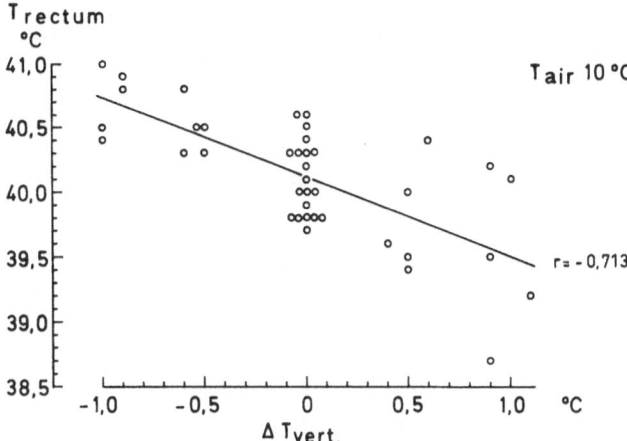

Fig. 31. Changes of rectal temperature induced in conscious pigeons by slight upward and downward displacements of spinal cord temperature, results obtained from 5 animals; normal core temperature 40.1° C. (RAUTENBERG, 1967; in: THAUER and SIMON, 1972)

As further shown by Fig. 29, BRÜCK and SCHWENNICKE (1971) have demonstrated a narrow, continuous distribution of the combinations of hypothalamic and subcutaneous temperatures at the threshold of non-shivering thermogenesis along a hyperbolic curve, in which hypothalamic temperature ranged from 34 to 41° C. Accordingly, the hypothalamic temperature sensor of the guinea pig is likewise continuously operating from the hypo- to the hyperthermic range of core temperatures.

2.2.3. Shifts of Core Temperature

As pointed out before, changes of core temperature induced by selective stimulation of a temperature sensor do not give a true estimate of the open loop gain of the stimulated feedback loop. However, according to the concept of multiple parallel inputs, at least minute opposing core temperature changes should be expected in response to slight upward and downward displacements of sensor temperature if this sensor continuously controls the thermoregulatory effectors. In the pigeon, RAUTENBERG (1969) could demonstrate a linear negative correlation between rectal temperature and spinal cord temperature for spinal temperature displacements within 1° C above and below normal core temperature (Fig. 31). JESSEN and CLOUGH (1973b) were able to show in goats that upward temperature displacements in the spinal canal of presumably only a few tenths of a degree induced measurable opposing changes of extraspinal core temperature. For downward displacements of spinal cord temperature, the significance of their observations is limited by the fact that spinal cold sensitivity is little developed in this species. Nevertheless, definite opposing changes of extraspinal core temperature could be demonstrated in response to slight downward displacements of spinal cord temperature under external cold conditions. At warm ambient temperatures, this effect was likewise detectable though less pronounced.

In the same manner, the hypothalamic temperature sensor of goats has been investigated by Jessen and Clough (1973a). Its continuous function in the control of the thermoregulatory effectors at normal core temperature could likewise be confirmed.

3. Characteristics of the Temperature Regulation System

If one would try to gather under one common heading the insights into the function of the temperature regulation system which have been gained from the reviewed investigations, one would, in the author's opinion, arrive at the statement that *multiplicity* as a principle of design in biological control systems (Jones, 1969) has been amply confirmed for presumably any function of the thermoregulatory control system. Multiplicity and redundancy of inputs, feedback loops and effector systems have been emphasized in recent discussions about the characteristics of the temperature regulation system (Brown and Brengelmann, 1970; Hensel, 1973). Recent models of temperature regulation have taken account of this idea (Bligh, 1972; Mitchell et al., 1972). Experimental support has been presented in this report that not only the sensor and effector functions but also the controller functions in temperature regulation may be multiplexly represented. The tentative diagram of Fig. 32 tries to comprehend these characteristics and to consider in addition some further aspects of the biological temperature regulation system in a descriptive, generalized form.

3.1. Multiplicity of Sensors

In retrospect, evaluation of the spinal cord as a temperature sensor appears as the first successful attempt to challenge the classical view of hypothalamus as the only central thermoreceptive region (Benzinger et al., 1961). In the meantime, additional observations concerning thermoregulatory reactions in response to thermal stimulation of the brain stem outside of the hypothalamic region have supported the suggestion that the anatomical distribution of central nervous thermoregulatory mechanisms might reflect the segmental organization of the central nervous axis (Simon, 1967). With respect to the midbrain, autonomic cold defence reactions were observed during local cooling (Cabanac and Hardy, 1969; Hardy, 1969) in the rabbit, while no substantial thermoregulatory responses could be elicited in monkeys (Adair and Stitt, 1971). However, Lipton (1971, 1973) and Chai and Lin (1972, 1973) have unmistakably demonstrated the appropriate activation of autonomic and behavioral thermoregulatory effector mechanisms by modifying the temperature of the medulla oblongata. In accordance with these demonstrations of a specific thermosensitivity in the lower brain stem, thermosensitive neurons were found to be distributed from the preoptic anterior hypothalamus (Nakayama et al., 1961, 1963) over the tuberal (Edinger and Eisenman, 1970), posterior (Nutik, 1971) and caudal hypothalamus (Wünnenberg and Hardy, 1972) to the midbrain (Nakayama and Hardy, 1969) and to the medulla (Boulant and Hardy, 1972).

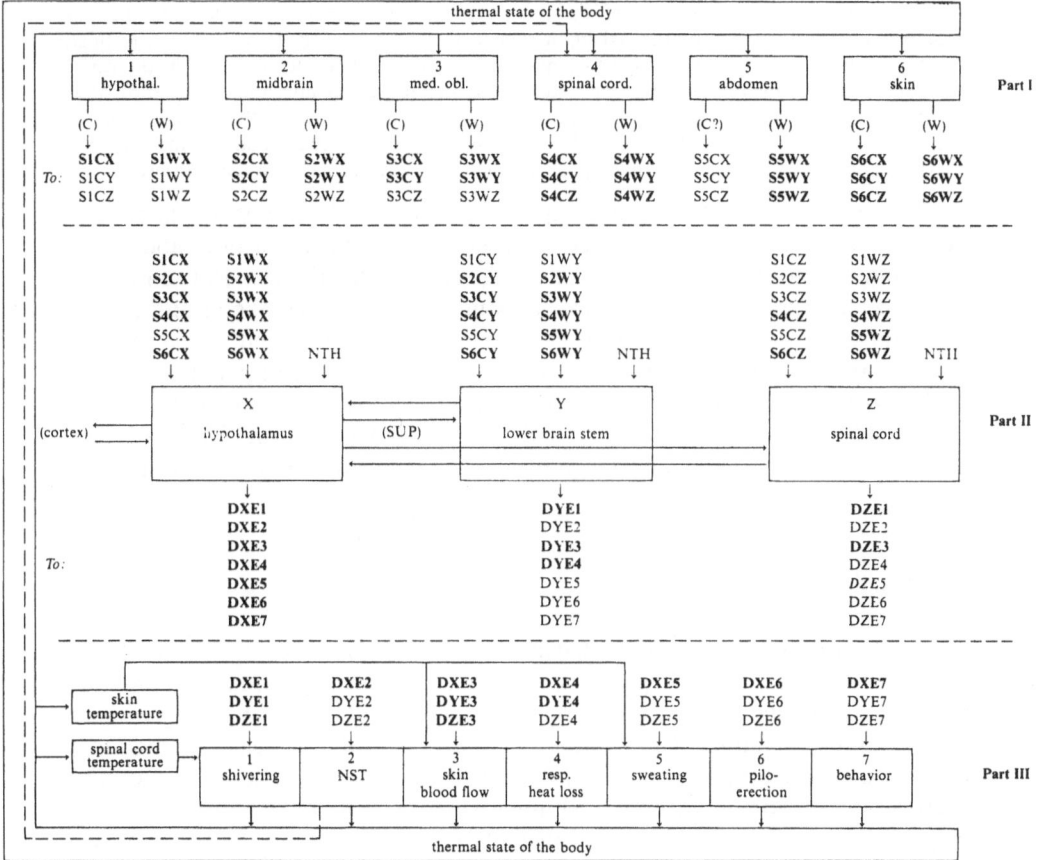

Fig. 32. Block diagram describing the temperature regulation system with multiple sensors (part I), multiple controllers (part II) and multiple effectors (part III). C: *cold* signals; W: *warm* signals; S...: thermal *sensory* inputs; NTH: *non-thermal* inputs; SUP: control loops constituting the *supervisor* function of the hypothalamus; D...: outputs *driving* the effectors. Bold letters: connections between sensors and controllers and between controllers and effectors which are experimentally indicated or are likely to exist for anatomical reasons. It is assumed that each sensor-effector connection may be described by: $R - R_0 = \alpha_R (T_{sensor} - T_{set(R)})$; $R \geqq R_0$, and that α_R and $T_{set(R)}$ are functions of the remaining inputs. Further details see text

RAWSON and QUICK (1970, 1971a, b, 1972) have further shown that central thermal inputs into the temperature regulation system originate from deep thermo-receptors outside of the central nervous system. By observing the effects of intra-abdominal heating in sheep with and without splanchnicotomy these authors demonstrated the existence of warm sensitive structures in the abdominal region. RIEDEL et al. (1973) have confirmed these findings in the rabbit and have further revealed a linear increase of temperature dependent afferent splanchnic activity with rising intra-abdominal temperature in the temperature range above 36° C.

The quantitative contributions of the mesencephalic, medullary and abdominal thermo-receptive structures to the total signal input remain to be determined. It is, however, justified to assume that these sensors provide informations about deep body temperature in the range of natural core temperature variations. Regarding the mesencephalic and rhombencephalic temperature detectors, their characteristics are probably quite similar to those located in the hypothalamus and in the spinal cord. With respect to the intra-abdominal thermoreceptors, RIEDEL et al. (1973) have shown that signals are generated at normal core temperature.

As a consequence, if the intire signal input into the thermoregulatory control system is to be described, mesencephalic, medullary, spinal and intra-abdominal temperature sensors also have to be considered besides the classical sensors of the skin and hypothalamus. It may be assumed that the peripheral and central temperature sensors, perhaps with the exception of the intra-abdominal thermo-sensitive region, provide both cold signals and warm signals. In part I of Fig. 32, each of the various possible input channels is labelled by a particular combination of symbols. It is assumed that the cold signals and the warm signals of each sensor are qualitatively equivalent but may differ in quantitative respects.

3.2. Multiplicity of Controllers

After the discovery of extrahypothalamic thermoregulatory control functions in spinalized animals, thermoregulatory responses have also been found in anesthe-tized mesencephalic preparations. SIMON et al. (1968) observed shivering in mid-brain transected rabbits. CHAI and LIN (1973) were able to elicit in addition appropriate vasoconstrictor and respiratory reactions by peripheral and central thermal stimulation in decerebrated unanesthetized rabbits. This finding is of particular interest as it shows that medullary structures are able to transform temperature signals into appropriate respiratory responses. The observations in spinalized and mesencephalic animals have, thus, provided substantial evidence for the concept that control functions in temperature regulation may be exerted by integrating mechanisms widespread in the central nervous system. Several recent observations in hypothalamic-lesioned rats (LIPTON, 1968, 1973; CARLISLE, 1969; SATINOFF and RUTSTEIN, 1970; SATINOFF and SHAN, 1971; TOTH, 1973) and bats (KLUGER and HEATH, 1971 a, b) have revealed a considerable degree of thermoregulatory capabilities or only differential impairments of thermo-regulatory responses in these animals. KLUGER and HEATH (1971 b) have concluded that the preoptic-anterior hypothalamus has little thermal and integrative functions in the bat, a conclusion which implicitly points to the importance of extra-hypothalamic thermoregulatory control functions in this species. LIPTON (1973) has concluded from his observations in hypothalamic-lesioned rats that the medulla not only "provides subsidiary temperature information to the preoptic anterior hypothalamic region" but also "makes up a secondary temperature control". This view is in full accordance with the conclusions drawn from the investigations in spinalized and decerebrated animals. Accordingly, the tempera-ture regulation system as a whole may be characterized as being composed of multiple controllers which may in principle function independently of each other. Since thermoregulatory responses were found to be mediated by spinal and medullary structures, controller functions are tentatively ascribed to the hypothalamus, the lower brain stem and the spinal cord in Part II of Fig. 32[1].

A further subdivision of the controlling mechanisms according to the hypothesis of the segmental representation of thermoregulatory functions (SIMON, 1967) appears to have little meaning at this time, since the hypothetical segmental thermoregulatory functions have not yet been experimentally elucidated. The diagram also does not take into account the possibility that the nervous mechanisms controlling the diverse effector systems might be anatomically separate to some degree. SATINOFF and RUTSTEIN (1970) and SATINOFF and SHAN (1971) have shown that autonomic and behavioral thermoregulatory responses can be separately excluded by placing lesions in the medial or lateral parts of the hypothalamus. An anatomical separation of heat accumulating and heat dissipating control mechanisms in the hypothalamus according to the concept of MEYER (1913) was repeatedly discussed in the past by KELLER (1935, 1963) and RANSON and MAGOUN (1939), to mention a few, on the basis of results obtained from lesioned animals. The effects of hypothalamic lesions are, however, difficult to interpret, since the loss of control functions, the loss of sensory functions and the interruption of pathways may be involved. The arguments in favour of and against the assumption that neutral structures controlling the various effector systems may be anatomically separate have recently been discussed by BLIGH (1973).

Two questions follow from the assumption of multiple controllers in temperature regulation: firstly, to what degree does each of these multiple controllers contribute to temperature regulation in intact animals and, secondly, in what way do these controllers cooperate in order to account for the high degree of precision and reliability in temperature regulation?

With respect to the first question, it would correspond to the phylogenetic principle of "cephalization" that thermoregulatory mechanisms distributed along the whole central nervous axis have become most concentrated in its rostral parts (THAUER, 1973). It has already been pointed out that the efficiency of thermoregulatory functions in spinalized animals is low, though not negligible as demonstrated previously by THAUER (1935). Since spinal temperature regulation can be demonstrated only after spinal transection, its spectrum of effector responses is necessarily reduced, because the supraspinal efferent contributions are excluded. The observations of LIPTON (1973) suggest that the thermoregulatory capabilities of the central nervous axis, in which only the hypothalamic control mechanisms are severed, may be considerably greater than those of the isolated spinal cord alone. The reduced effectivity and accuracy of extrahypothalamic, "coarse control" mechanisms (BLIGH, 1973) may be due to their low gain or to deficient controller-effector connections rather than to an inadequate

---

1 Substantial further support to this concept of multiple representation of thermoregulatory controller functions at different levels of the central nervous axis has been added by two recent publications. W. W. CHAMBERS, M. S. SEIGEL, J. C. LIU, and C. N. LIU [Thermoregulatory responses of decerebrate and spinal cats — Exp. Neurol. 42, 282–299 (1974)] have demonstrated a considerable degree of thermoregulatory cold defence capabilities in low decerebrate cats. By proposing inhibitory influences of higher midbrain structures on the thermoregulatory controlling mechanisms in the lower brain stem and spinal cord these authors offer an explanation for the earlier controversial findings concerning the thermoregulatory capabilities of adult animals deprived of the hypothalamic thermoregulation center. K. E. BIGNALL and L. SCHRAMM [Behavior of chronically decerebrated kittens — Exp. Neurol. 42, 519–531 (1974)] have been able to demonstrate a nearly normal development of thermoregulation and of a variety of behavioral reactions in kittens decerebrated one week after birth. Their findings suggest important contributions of mesencephalic and more caudal central nervous structures to behavioral and regulatory processes which were thought to be organized primarily in the prosencephalon.

thermal sensory input. Central extrahypothalamic as well as hypothalamic and cutaneous temperature sensors are operating continuously and their temperature signals may be regarded as qualitatively equivalent.

It is, however, impossible for different reasons to draw definite conclusions on the role of extrahypothalamic thermoregulatory functions in intact animals from the observations in lesioned animals. If hypothalamic lesions or brain stem transections have been performed, additional impairment of nervous functions not specifically related to temperature regulation (disturbances of endocrine and γ-motor functions) may considerably reduce the effectivity of the lower thermoregulatory mechanisms. On the other hand, the ability of the central nervous system to compensate for anatomical defects by reorganizing central nervous functions has to be taken into account. The gradual improvement of cutaneous vasomotor responses to cold stimulation in chronically spinalized animals (WALTHER et al., 1971a) and, consequently, the gradual improvement of cold tolerance (THAUER, 1935) reflects this ability of central nervous structures. HAMILTON and BROBECK (1964) and CHOWERS et al. (1972) have demonstrated the gradual restitution of cold tolerance in rats with extensive hypothalamic lesions or, respectively, in hypothalamic-disconnected rats.

With respect to the second question concerning the cooperation of the multiple controllers, the fact has to be considered that the thermoregulatory control functions of intact animals and of man have been successfully described by models proposing only one central controller. This might, of course, be due to a preponderance of the hypothalamic control mechanisms so great as to account entirely for the regulation of body temperature. This possibility may be disregarded, though not denied, in the present considerations which try to meet the fact that extrahypothalamic controlling systems do exist. If these are assumed to take part in temperature regulation, whatever their quantitative contribution may be, a high degree of integration must be required. This is especially indicated by the finding that a common error signal appears to be generated under virtually any condition of selective and combined, equidirectional and opposing thermal stimulations of various temperature sensors. Reports about temporary, simultaneous activations of opposing effector mechanisms under particular experimental conditions are rare (HAMMEL et al., 1963a; CABANAC, 1970, 1972) and might be related to the individuality of the experimental animal (CABANAC, 1972). In an attempt to take into account both the independent and the integrated functions of the multiple controllers, parallel as well as hierarchical arrangements have been proposed in Part II of Fig. 32.

On the one hand, spinal, medullary and hypothalamic control centers are supposed to be arranged in parallel, each center receiving warm and cold input signals from various sensors and providing output signals for various effectors, of course within the limits drawn by the anatomical interconnections. In the block diagram the theoretically possible input and output connections are indicated by the identical combinations of symbols in part I and part II and, respectively, in part II and part III. They are supposed to differ only in quantitative respects. This meshing of the diverse control loops according to the principles of divergence and convergence of signal pathways might constitute a multiplexed system of feedback mechanisms.

On the other hand, integrating mechanisms are required which supervise, in a sense, the parallel feedback systems depicted in the block diagram. This integrating or supervisor function which is indicated by the horizontal arrows in part II of the diagram may be compared with a multilevel system of control. Since the highest degree of precision in temperature regulation is found when hypothalamic functions are all intact, the role of the supervisor has to be ascribed to the hypothalamus. The "switching function" of the head core temperature proposed by STOLWIJK and HARDY (1966) may be regarded as an analogue to this supervisor function of the hypothalamus. According to HARDY (1973) the role of the caudal hypothalamus in temperature regulation might be critical in that integration is *completed* at this level.

Part II of the block diagram further takes into account that temperature independent factors such as exercise, pyrogens, dehydratation, circadian rhythms etc. influence the input-output relations of the temperature regulation system. These factors have been lumped together under the term "non-thermal input". This term also comprises those biological mechanisms by which the *setpoint* of the controlling system is established. Since in a multiple sensor system all thermal inputs are regarded as providing informations about the controlled variable, the setpoint is assumed to be determined only by non-thermal factors (HENSEL, 1973).

The thermal setpoint has been a matter of controversy in the discussions about the principles according to which the temperature regulation system works. Its existence has been intuitively proposed since the first attempts to describe the phenomenon of homeothermia and its disturbance (LIEBERMEISTER, 1875). HAMMEL et al. (1963a) have inferred that a reference signal is fed into the central nervous controller by special thermoinsensitive neurons, the low-Q_{10}-units. Others (e.g. MITCHELL et al., 1972) have regarded the reference signal as an abstract entity which only defines the signal input at "zero" output as in technical control systems. It is not intended to enter the extensive discussion about the reference signal in temperature regulation, because the investigations reported in this review do not seem to have added substantial arguments in favour of one or the other concept. However, the fact ought to be considered that the central neurons involved in the control of body temperature are obviously influenced by various factors which are temperature independent in comparison to the thermoreceptive neural mechanisms. For instance, biogenic amines (BLIGH et al., 1971; ZEISBERGER and BRÜCK, 1971a, b) and Na^+- and Ca^{++}-ions (FELDBERG et al., 1970; MYERS and VEALE, 1970; MYERS and YAKSH, 1971) have been shown to differentially affect the hypothalamic thermoregulatory mechanisms when their concentrations are locally changed. The existence of a multitude of temperature independent factors which are relevant in temperature regulation raises the suggestion that the discussion about the reference signal might well turn out to be a matter of terminology concerning the problem of what may be called a "signal" in the temperature regulation system.

3.3. Multiplicity of Effectors

This well-known characteristic may be regarded as the most telling example for the multiple representation of functions in the biological temperature regulation system. The full spectrum of effector mechanisms has been known for a long time and the rules according to which skin and core temperatures interact in controlling their activities have been amply investigated. The experimental results reported in the first chapter of this review have shown that the spinal temperature sensor appropriately activates or inhibits most of the multiple effectors, perhaps with the exception of non-shivering thermogenesis. It is tentatively assumed in Part III of the block diagram that each of the multiple effectors may be appropriately influenced by each of the controllers. In this way, a multitude of possible sensor-effector connections might exist.

There have been reports in the literature demonstrating that thermal inputs from the skin and core may not be equally effective in driving or inhibiting the various thermoregulatory effectors. This has been explained by different densities of cold and warm receptors in the skin and core (BENZINGER, 1969). In addition, the comparative analysis of the effects of cutaneous, hypothalamic and spinal temperature displacements on different thermoregulatory effectors has raised the suggestion that there might exist what may be called a "predilection" of the cold signals or the warm signals of certain sensors for certain effectors.

Investigations of BRÜCK and WÜNNENBERG (1970) and of BRÜCK and SCHWENNICKE (1971) in guinea pigs have shown that the hypothalamic warm receptors are able to inhibit non-shivering thermogenesis but not cold shivering. Conversely, spinal warm stimuli inhibit cold shivering but do not affect non-shivering thermogenesis (WÜNNENBERG and BRÜCK, 1968b). Interestingly, the warm receptors of both sensors are about equally effective in stimulating thermal tachypnea in this species (HERRMANN, 1972). The observations of RAUTENBERG et al. (1972) in pigeons have indicated that displacements of brain and spinal cord temperature are equally effective in modifying skin blood flow. Spinal cord temperature controls shivering and panting in the pigeon while brain temperature has little influence on panting and no influence at all on shivering. According to JESSEN and SIMON (1971) the cold signal inputs from both the spinal cord and the hypothalamus appeared to have a much smaller influence on panting in dogs than the warm signals originating from the same regions. No such difference was found with regard to cold shivering. A discrepancy between the presence of autonomic thermoregulatory reactions and the absence of behavioral adjustments in response to spinal cooling was found in dogs (SPECTOR and CORMARÈCHE, 1971; CORMARÈCHE-LEYDIER and CABANAC, 1973). The investigations of CARLISLE and INGRAM (1973b) in pigs seem to contribute a further example: comparing the influences of spinal and hypothalamic thermal stimulations on thermoregulatory behavior both postural and operant behavior could be influenced by thermal stimulation of both sensors, but postural seemed behavior to be preferably controlled by spinal cord temperature, while operant behavior was dominated by hypothalamic temperature. The described preferential and weakly indicated connections between the cold or warm signal inputs of diverse sensors and the diverse effectors might reflect different gains of the various possible feedback loops converging upon the single effectors. Despite these quantitatively different input-output connections, the high degree of integration between the controlling mechanisms should guarantee that the activity of the temperature regulation system as a whole is directed towards either heat accumulation or heat dissipation. At least, the occurrence of simultaneous activations of opposing effector mechanisms appears to be largely prevented.

Finally, local temperature effects on several effector mechanisms are considered in Part III of the block diagram of Fig. 32. The modulating effects of local skin temperature on the vasomotor and sudomotor responses have been taken into account in accordance with HENSEL (1973). Local temperature effects on spinal motoneurons (KLUSSMANN and PIERAU, 1972) which may modulate the shivering response (see Paragraph 1.4.3.) have further been considered. The local temperature effects of non-shivering thermogenesis (NST) on the spinal temperature sensor (BRÜCK and WÜNNENBERG, 1970) may be regarded as a particular pathway within the general feedback channel established by the thermal state of the body.

Acknowledgement is given to the editors of the journals and books from which figures have been reproduced in this article.

The permission of the authors for publication of their figures in original or redrawn form is gratefully acknowledged.

Journals

American Journal of Physiology (American Physiological Society, Bethesda): Figs. 15, 30.
Experientia (Basel) (Birkhäuser Verlag, Basel): Figs. 2, 6.
International Journal of Biometeorology (Swets and Zeitlinger, N.V., Amsterdam): Fig. 21.
Journal de Physiologie (Paris) (Masson et Cie., Paris): Figs. 22, 24, 25, 26, 28.
Journal of Applied Physiolgy (American Physiological Society, Bethesda): Fig. 23.
Pflügers Archiv (Springer-Verlag, Berlin): Figs. 5, 7, 11, 12, 14, 16, 17, 19, 27.
The Journal of Physiology (London) (Cambridge University Press, London): Fig. 10.

Books

S. Itoh, K. Ogata and H. Yoshimura: *Advances in Climatic Physiology* (Igaku Shoin, Ltd., Tokyo 1972): Figs. 4, 13.

J. Bligh and R. Moore: *Essays on Temperature Regulation* (North-Holland Publ. Comp., Amsterdam 1972): Fig. 20.

J. D. Hardy, A. Ph. Gagge and J. A. J. Stolwijk: *Physiological and Behavioral Temperature Regulation* (Charles C. Thomas Publ., Springfield 1970): Figs. 3, 9.

References

Adair, E. R.: Control of thermoregulatory behavior by brief displacements of hypothalamic temperature. Psychon. Sci. **20**, 11–13 (1970).

Adair, E. R.: Behavioral temperature regulation in the squirrel monkey: some limits of hypothalamic control. Medical Primatology, Proc. 2nd Conf. exp. Med. Surg. Primates, New York 1969, pp. 462–470. Basel: S. Karger 1971.

Adair, E. R., Casby, J. U. Stolwijk, J. A. J.: Behavioral temperature regulation in the squirrel monkey: changes induced by shifts in hypothalamic temperature. J. comp. physiol. Psychol. **72**, 17–27 (1970).

Adair, E. R., Stitt, J. T.: Behavioral temperature regulation in the squirrel monkey: effects of midbrain temperature displacements. J. Physiol. (Paris) **63**, 191–194 (1971).

Andersen, H. T., Andersson, B., Gale, C. C.: Central control of cold defence mechanisms and the release of "endopyrogen" in the goat. Acta physiol. scand. **54**, 159–174 (1962).

Andersson, B., Ekman, L., Gale, C. C., Sundsten, J. W.: Control of thyrotrophic hormone (TSH) secretion by the "heat loss center". Acta physiol. scand. **59**, 12–33 (1963a)·

Andersson, B., Gale, C. C., Sundsten, J. W.: Effects of chronic central cooling on alimentation and thermoregulation. Acta physiol. scand. **55**, 177–188 (1962).

Andersson, B., Gale, C. C., Sundsten, J. W.: The relationship between body temperature and food and water intake. Y. Zotterman: Olfaction and taste, pp. 361–375. Oxford: Pergamon Press 1963b.

Andersson, B., Larsson, B.: Influence of local temperature changes in the preoptic area and rostral hypothalamus on the regulation of food and water intake. Acta physiol. scand. **52**, 75–89 (1961).

Baldwin, B. A., Ingram, D. L.: The effect of heating and cooling the hypothalamus on behavioural thermoregulation in the pig. J. Physiol. (Lond.) **191**, 375–392 (1967).

Barbour, H. G.: Die Wirkung unmittelbarer Erwärmung und Abkühlung der Wärmezentra auf die Körpertemperatur. Naunyn-Schmiedeberg's Archiv **70**, 1–26 (1912).

Barker, J. L., Carpenter, D. O.: Thermosensitivity of neurons in the sensorimotor cortex of the cat. Science **169**, 597–598 (1970).

Barron, D. H., Matthews, B. H. C.: Intermittent conduction in the spinal cord. J. Physiol. (Lond.) **85**, 73–103 (1935).

Barron, D. H., Matthews, B. H. C.: Dorsal root potentials. J. Physiol. (Lond.) **94**, 27P–29P (1938).

Beaton, L. E., McKinley, W. A., Berry, C. M., Ranson, S. W.: Localization of cerebral center activating heat-loss mechanisms in monkeys. J. Neurophysiol. **4**, 478–485 (1941).

Beckman, A. L., Amini-Kormi, L.: Responses of rat hypothalamic thermosensitive cells to iontophoretically applied sodium salicylate. Int. J. Biometeor. **16**, Suppl.: Biometeorology **5**, 37 (1972).

Beckman, A. L., Eisenman, J. S.: Microelectrophoresis of biogenic amines on hypothalamic thermosensitive cells. Science **170**, 334–336 (1970).

Benzinger, T. H.: Heat regulation: homeostasis of central temperature in man. Physiol. Rev. **49**, 671–759 (1969).

Benzinger, T. H., Pratt, A. W., Kitzinger, Ch.: The thermostatic control of human metabolic heat production. Natl. Acad. Sci. **47**, 730–739 (1961).

Bethe, A.: Plastizität und Zentrenlehre. In: A. Bethe, G. v. Bergmann, G. Embden und A. El-
    linger, Hdb. norm. path. Physiol. 15 II, 1175–1220 (1931).
Betz, E., Brück, K., Hensel, H., Járai, J., Malan, A.: Verhalten des Energieumsatzes bei
    umschriebener Hypothalamuskühlung an der wachen Katze. Pflügers Arch. 272, 76–77 (1960).
Birzis, L., Hemingway, A.: Descending brain stem connections controlling shivering in cat.
    J. Neurophysiol. 19, 37–43 (1956).
Birzis, L., Hemingway, A.: Efferent brain discharge during shivering. J. Neurophysiol. 20,
    156–166 (1957).
Blatteis, C. M.: Afferent initiation of shivering. Amer. J. Physiol. 199, 697–700 (1960).
Bligh, J.: Possible temperature-sensitive elements in or near the vena cava of sheep. J. Physiol.
    (Lond.) 159, 85 P (1961).
Bligh, J.: The thermosensitivity of the hypothalamus and thermoregulation in mammals.
    Biol. Rev. 41, 317–367 (1966).
Bligh, J.: Neuronal models of mammalian temperature regulation. In: J. Bligh and R. Moore,
    Essays on Temperature Regulation, pp. 105–120. Amsterdam: North Holland Publ. Comp.
    1972.
Bligh, J.: Temperature regulation in mammals and other vertebrates. Amsterdam: North
    Holland Publ. Comp. 1973.
Bligh, J., Cottle, W. H., Maskrey, M.: Influence of ambient temperature on the thermo-
    regulatory responses to 5-hydroxytryptamine, noradrenaline and acetylcholine injected into
    the lateral cerebral ventricles of sheep, goats, and rabbits. J. Physiol. (Lond.) 212, 377–392
    (1971).
Bligh, J., Johnson, K. G.: Glossary of terms for thermal physiology. J. Appl. Physiol. 35,
    941–960 (1973).
Bost, J., Dorleac, E.: Action du froid sur le taux plasmatique des acides gras non estérifiés
    (FFA) chez le mouton. C.R. Séances Soc. Biol. (Paris) 159, 2209–2212 (1965).
Boulant, J. A., Hardy, J. D.: Hypothalamic and medullary neuronal responses to deep-body,
    peripheral and local temperatures. Internat. J. Biometeor. 16, Suppl.: Biometeorology 5,
    38–39 (1972).
Brendel, W.: Die Bedeutung der Hirntemperatur für die Kältegegenregulation. II. Der Ein-
    fluß der Hirntemperatur auf den respiratorischen Stoffwechsel des Hundes unter Kälte-
    belastung. Pflügers Arch. 270, 628–647 (1960).
Brooks, C. McC., Koizumi, K., Malcolm, J. L.: Effects of changes in temperature on reactions
    of spinal cord. J. Neurophysiol. 18, 205–216 (1955).
Brown, A. C., Brengelmann, G. L.: The interaction of peripheral and central inputs in the
    temperature regulation system. In: J. D. Hardy, A. Ph. Gagge and J. A. J. Stolwijk,
    Physiological and Behavioral Temperature Regulation, pp. 684–702. Springfield: Charles C.
    Thomas Publ. 1970.
Brück, K., Gallmeier, H., Wünnenberg, W.: Einfluß der Temperatur des vorderen Hypo-
    thalamus auf die zitterfreie Thermogenese des Meerschweinchens. Pflügers Arch. 294, R 86
    (1967).
Brück, K., Schwennicke, H. P.: Interaction of superficial and hypothalamic thermosensitive
    structures in the control of non-shivering thermogenesis. Int. J. Biometeor. 15, 156–161 (1971).
Brück, K., Wünnenberg, W.: Beziehung zwischen Thermogenese im „braunen" Fettgewebe,
    Temperatur im cervicalen Anteil des Vertebralkanals und Kältezittern. Pflügers Arch. 290,
    167–183 (1966).
Brück, K., Wünnenberg, W.: Die Steuerung des Kältezitterns beim Meerschweinchen. Pflügers
    Arch. 293, 215–225 (1967a).
Brück, K., Wünnenberg, W.: Eine kälteadaptative Modifikation: Senkung der Schwellen-
    temperaturen für Kältezittern. Pflügers Arch. 293, 226–235 (1967b).
Brück, K., Wünnenberg, W.: "Meshed" control of two effector systems: nonshivering and
    shivering thermogenesis. In: J. D. Hardy, A. Ph. Gagge and J. A. J. Stolwijk, Physio-
    logical and Behavioral Temperature Regulation, pp. 562–580. Springfield: Charles C. Thomas
    Publ. 1970.
Brück, K., Wünnenberg, W., Gallmeier, H., Ziehm, B.: Shift of threshold temperature for
    shivering and heat polypnea as a mode of thermal adaptation. Pflügers Arch. 321, 159–172
    (1970).

BRÜCK, K., WÜNNENBERG, W., GALLMEIER, H., ZIEHM, B.: A mode of thermal adaptation: Shift of threshold temperatures for shivering and heat polypnea. J. Physiol. (Paris) **63**, 213–215 (1971).

BULLARD, R. W., BANERJEE, M. R., CHEN, F., ELIZONDO, R., MacINTYRE, B. A.: Skin temperature and thermoregulatory sweating: a control systems approach. In: J. D. HARDY, A. PH. GAGGE and J. A. J. STOLWIJK, Physiological and Behavioral Temperature Regulation, pp. 597–610. Springfield: Charles C. Thomas Publ. 1970.

CABANAC, M.: Interaction of cold and warm temperature signals in the brain stem. In: J. D. HARDY, A. PH. GAGGE and J. A. J. STOLWIJK, Physiological and Behavioral Temperature Regulation, pp. 549–561. Springfield: Charles C. Thomas Publ. 1970.

CABANAC, M.: Thermoregulatory behavior. In: J. BLIGH and R. MOORE, Essays on Temperature Regulation, pp. 19–36. Amsterdam: North Holland Publ. Comp. 1972.

CABANAC, M., HARDY, J. D.: Réponses unitaires et thermorégulatrices hors de réchauffments et réfroidissements localisés de la région préoptique et du mésencéphale chez le lapin. J. Physiol. (Paris) **61**, 331–347 (1969).

CABANAC, M., JEDDI, E.: Thermopreferendum et thermoregulation comportementale chez trois poikilothermes. Physiology and Behavior **7**, 375–380 (1971).

CABANAC, M., STOLWIJK, J. A. J., HARDY, J. D.: Effect of temperature and pyrogens on single-unit activity in the rabbit's brain stem. J. Appl. Physiol. **24**, 645–652 (1968).

CARLISLE, H. J.: Behavioural significance of hypothalamic temperature-sensitive cells. Nature **209**, 1324–1325 (1966).

CARLISLE, H. J.: Effect of preoptic and anterior hypothalamic lesions on behavioral thermoregulation in the cold. J. comp. physiol. Psych. **69**, 391–402 (1969).

CARLISLE, H. J., INGRAM, D. L.: The influence of body core temperature and peripheral temperatures on oxygen consumption in the pig. J. Physiol. (Lond.) **231**, 341–352 (1973a).

CARLISLE, H. J., INGRAM, D. L.: The effects of heating and cooling the spinal cord and hypothalamus on thermoregulatory behaviour in the pig. J. Physiol. (Lond.) **231**, 353–364 (1973b).

CHAI, C. Y., LIN, M. T.: Effects of heating and cooling the spinal cord and medulla oblongata on thermoregulation in monkeys. J. Physiol. (Lond.) **225**, 297–308 (1972).

CHAI, C. Y., LIN, M. T.: Effects of thermal stimulation of medulla oblongata and spinal cord on decerebrate rabbits. J. Physiol. (Lond.) **234**, 409–419 (1973).

CHATONNET, J., TANCHE, M.: Dissociation du frisson «central» et du frisson «réflexe» chez le chien à moelle détruite. J. Physiol. (Paris) **48**, 439–442 (1956).

CHATONNET, J., TANCHE, M.: Sur l'origine des réactions thermogénétiques déclenchées par ingestion d'eau glacée chez le chien. C.R. Soc. Biol. (Paris) **151**, 1533–1535 (1957a).

CHATONNET, J., TANCHE, M.: Arguments en faveur d'une régulation centrale de la thermogénèse chez le chien. J. Physiol. (Paris) **49**, 89–91 (1957b).

CHOWERS, I., CONFORTI, N., FELDMAN, S.: Body temperature and adrenal function in cold-exposed hypothalamic-disconnected rats. Amer. J. Physiol. **223**, 341–345 (1972).

CLOUGH, D. P., DARLING, K. F., FINDLAY, J. D., THOMPSON, G. E.: Cold sensitivity in the spinal cord of sheep. Pflügers Arch. **342**, 137–144 (1973).

CORBIT, J. D.: Behavioral regulation of hypothalamic temperature. Science **166**, 256–258 (1969).

CORBIT, J. D.: Behavioral regulation of body temperature. In: J. D. HARDY, A. PH. GAGGE and J. A. J. STOLWIJK, Physiological and Behavioral Temperature Regulation, pp. 777–801. Springfield: Charles C. Thomas Publ. 1970.

CORMARÈCHE-LEYDIER, M., CABANAC, M.: Influence de stimulations thermiques de la moelle épinière sur le comportement thermorégulateur du chien. Pflügers Arch. **341**, 313–324 (1973).

CRAWSHAW, L. I., HAMMEL, H. T.: Behavioral thermoregulation in two species of antarctic fish. Life Sci. **10**, Part I, 1009–1020 (1971).

CUNNINGHAM, D. J., STOLWIJK, J. A. J., MURAKAMI, N., Hardy, J. D.: Responses of neurons in the preoptic area to temperature. serotonin, and epinephrine. Amer. J. Physiol. **213**, 1570–1581 (1967).

DASTRE, A., MORAT, I. P.: Influence du sang asphyxique sur l'appareil nerveux de la circulation. Arch. Physiol. norm. path. **16**, 1–45 (1884).

DELIUS, W., HAGBARTH, K.-E., HONGELL, A., WALLIN, B. G.: Manœuvres affecting sympathetic outflow in human skin nerves. Acta physiol. scand. **84**, 177–186 (1972).

Donhoffer, Sz., Farkas, M., Haug-László, A., Járai, I., Szegvári, Gy.: Das Verhalten der Wärmeproduktion und der Körpertemperatur der Ratte bei lokaler Erwärmung und Kühlung des Gehirnes. Pflügers Arch. **268**, 273–280 (1959).

Downey, J. A., Mottram, R. F.: Effect of regional blood stream cooling in the conscious rabbit. Proc. 22. Internat. Physiol. Congr., Vol. II, Abstr. 485, Leiden 1962.

Duclaux, R., Fantino, M., Cabanac, M.: Comportement thermoregulateur chez Rana esculenta. Influence du réchauffement spinal. Pflügers Arch. **342**, 347–358 (1973).

Dworkin, S.: Observations on the central control of shivering and of heat regulation in the rabbit. Amer. J. Physiol. **93**, 227–244 (1930).

Edinger, H. M., Eisenman, J. S.: Thermosensitive neurons in tuberal and posterior hypothalamus of cats. Amer. J. Physiol. **219**, 1098–1103 (1970).

Eisenman, J. S.: Pyrogen-induced changes in the thermosensitivity of septal and preoptic neurons. Amer. J. Physiol. **216**, 330–334 (1969).

Eisenman, J. S.: Unit activity studies of thermoresponsive neurons. In: J. Bligh and R. Moore, Essays on Temperature Regulation, pp. 55–69. Amsterdam: North Holland Publ. Comp. 1972.

Eisenman, J. S., Jackson, D. C.: Thermal response patterns of septal and preoptic neurons in cats. Exptl. Neurol. **19**, 33–45 (1967).

Euler, C. v.: The gain of the hypothalamic temperature regulating mechanisms. Progr. in Brain Res. **5**, 127–131. Amsterdam: Elsevier, 1964.

Euler, C. v., Söderberg, U.: Co-ordinated changes in temperature thresholds for thermoregulatory reflexes. Acta physiol. scand. **42**, 112–129 (1958).

Feldberg, W. S.: The monoamines of the hypothalamus as mediators of temperature responses. In: J. D. Hardy, A. Ph. Gagge, and J. A. J. Stolwijk, Physiological and Behavioral Temperature Regulation, pp. 493–506. Springfield: Charles C. Thomas Publ. 1970.

Feldberg, W., Myers, R. D.: Effects on temperature of amines injected into the cerebral ventricles. A new concept of temperature regulation. J. Physiol. (Lond.) **173**, 226–237 (1964).

Feldberg, W., Myers, R. D., Veale, W. L.: Perfusion from cerebral ventricle to cisterna magna in the unanaesthetized cat. Effect of calcium on body temperature. J. Physiol. (Lond.) **207**, 403–416 (1970).

Freeman, W. J., Davis, D. D.: Effects on cats of conductive hypothalamic cooling. Amer. J. Physiol. **197**, 145–148 (1959).

Fusco, M. M., Hardy, J. D., Hammel, H. T.: Interaction of central and peripheral factors in physiological temperature regulation. Amer. J. Physiol. **200**, 572–580 (1961).

Gale, C. C., Mathews, M., Young, J.: Behavioral thermoregulatory responses to hypothalamic cooling and warming in the baboon. Physiol. Behav. **5**, 1–6 (1970).

Goltz, F., Ewald, J. R.: Der Hund mit verkürztem Rückenmark. Pflügers Arch. **63**, 362–400 (1896).

Grayson, J.: Vascular reactions in the human intestine. J. Physiol. (Lond.) **109**, 439–447 (1949).

Grundfest, H.: The augmentation of the motor root reflex discharge in the cooled spinal cord of the cat. Amer. J. Physiol. **133**, P 307 (1941).

Guieu, J. D., Hardy, J. D.: Effects of preoptic and spinal cord temperature in control of thermal polypnea. J. Appl. Physiol. **28**, 540–542 (1970a).

Guieu, J. D., Hardy, J. D.: Effects of heating and cooling of the spinal cord on preoptic unit activity. J. Appl. Physiol. **29**, 675–683 (1970b).

Guieu, J. D., Hardy, J. D.: Integrative activity of preoptic units. I. Response to local and peripheral temperature changes. J. Physiol. (Paris) **63**, 253–256 (1971).

Hales, J. R. S., Jessen, C.: Increase of cutaneous moisture loss caused by local heating of the spinal cord in the ox. J. Physiol. (Lond.) **204**, 40 P–42 P (1969).

Hallwachs, O., Hupfer, H., Thauer, R.: Die Bedeutung der tiefen Körpertemperatur für die Auslösung der chemischen Temperaturregulation. I. Kältezittern durch Senkung der tiefen Körpertemperatur bei konstanter, erhöhter Hauttemperatur. Pflügers Arch. **274**, 97–114 (1961a).

Hallwachs, O., Thauer, R., Usinger, W.: Die Bedeutung der tiefen Körpertemperatur für die Auslösung der chemischen Temperaturregulation. II. Kältezittern durch Senkung der tiefen Körpertemperatur bei konstanter, erhöhter Haut- und Hirntemperatur. Pflügers Arch. **274**, 115–124 (1961b).

HAMILTON, C. L.: Food and temperature. Hdb. Physiol. Sect. 6, Vol. I, 303–317, Amer. Physiol. Soc., Washington D.C. 1967.

HAMILTON, C. L., BROBECK, J. R.: Food intake and temperature regulation in rats with rostral hypothalamic lesions. Amer. J. Physiol. **207**, 291–297 (1964).

HAMILTON, C. L., CIACCIA, P. J.: Hypothalamus, temperature regulation, and feeding in the rat. Amer. J. Physiol. **221**, 800–807 (1971).

HAMMEL, H. T.: Neurons and temperature regulation. In: W. S. YAMAMOTO and J. R. BROBECK, Physiological controls and regulations, pp. 71–97. Philadelphia: Saunders 1965.

HAMMEL, H. T.: Regulation of internal body temperature. Annual Rev. Physiol. **30**, 641–710 (1968).

HAMMEL, H. T.: Concept of the adjustable set temperature. In: J. D. HARDY, A. PH. GAGGE, and J. A. J. STOLWIJK, Physiological and Behavioral Temperature Regulation, pp. 676–683. Springfield: Charles C. Thomas Publ. 1970.

HAMMEL, H. T.: The set-point in temperature regulation: analogy or reality. In: J. BLIGH and R. MOORE, Essays on Temperature Regulation, pp. 121–137. Amsterdam: North Holland Publ. Comp. 1972.

HAMMEL, H. T., CRAWSHAW, L. I., CABANAC, H. P.: The activation of behavioral responses in the regulation of body temperature in vertebrates. In: E. SCHÖNBAUM and P. LOMAX, The pharmacology of thermoregulation, pp. 124–141. Basel: S. Karger 1973.

HAMMEL, H. T., HARDY, J. D., FUSCO, M. M.: Thermoregulatory responses to hypothalamic cooling in unanesthetized dogs. Amer. J. Physiol. **198**, 481–486 (1960).

HAMMEL, H. T., JACKSON, D. C., STOLWIJK, J. A. J., Hardy, J. D., STRÖMME, J. B.: Temperature regulation by hypothalamic proportional control with an adjustable set point. J. Appl. Physiol. **18**, 1146–1154 (1963a).

HAMMEL, H. T., SHARP, F.: Thermoregulatory salivation in the running dog in response to preoptic heating and cooling. J. Physiol. (Paris) **63**, 260–263 (1971).

HAMMEL, H. T., STRÖMME, S., CORNEW, R. W.: Proportionality constant for hypothalamic proportional control of metabolism in unanesthetized dog. Life Sci. **12**, 933–947 (1963b).

HARDY, J. D.: Physiology of temperature regulation. Physiol. Rev. **41**, 521–606 (1961).

HARDY, J. D.: The "setpoint" concept in physiological temperature regulation. In: W. S. YAMAMOTO and J. R. BROBECK, Physiological controls and regulations, pp. 98–116. Philadelphia: Saunders 1965.

HARDY, J. D.: Thermoregulatory responses to temperature changes in the midbrain of the rabbit. (Abstr. 2539). Fed. Proc. **28**, 713 (1969).

HARDY, J. D.: Models of temperature regulation—a review. In: J. BLIGH and R. MOORE, Essays on Temperature Regulation, pp. 163–186. Amsterdam: North Holland Publ. Comp. 1972a.

HARDY, J. D.: Peripheral inputs to the central regulator for body temperature. In: S. ITOH, K. OGATA and H. YOSHIMURA, Advances in climatic physiology, pp. 3–21. Tokyo: Igaku Shoin Ltd. 1972b.

HARDY, J. D.: Posterior hypothalamus and the regulation of body temperature. Fed. Proc. **32**, 1564–1571 (1973).

HARDY, J. D., GUIEU, J. D.: Integrative activity of preoptic units. II. Hypothetical network. J. Physiol. (Paris) **63**, 264–267 (1971).

HASHIMOTO, M.: Fieberstudien. II. Mitteilung. Über den Einfluß unmittelbarer Erwärmung und Abkühlung des Wärmezentrums auf die Temperaturwirkungen von verschiedenen pyrogenen und antipyretischen Substanzen. Naunyn-Schmiedeberg's Arch. **78**, 394–444 (1915).

HELLON, R. F.: Hypothalamic neurons responding to changes in hypothalamic and ambient temperatures. In: J. D. HARDY, A. PH. GAGGE, and J. A. J. STOLWIJK, Physiological and Behavioral Temperature Regulation, pp. 463–471. Springfield: Charles C. Thomas Publ. 1970a.

HELLON, R. F.: The stimulation of hypothalamic neurones by changes in ambient temperature. Pflügers Arch. **321**, 56–66 (1970b).

HELLON, R. F.: Temperature-sensitive neurons in the brain stem: their responses to brain temperature at different ambient temperatures. Pflügers Arch. **335**, 323–334 (1972a).

HELLON, R. F.: Central thermoreceptors and thermoregulation. In: E. NEIL, Hdb. of Sensory Physiology, Vol. III/1, pp. 161–186. Berlin-Heidelberg-New York: Springer 1972b.

Hellon, R. F.: Central transmitters and thermoregulation. In: J. Bligh and R. Moore, Essays on Temperature Regulation, pp. 71–85. Amsterdam: North Holland Publ. Comp. 1972c.

Hellström, B., Hammel, H. T.: Some characteristics of temperature regulation in the un-anesthetized dog. Amer. J. Physiol. 213, 547–556 (1967).

Hemingway, A., Rasmussen, T., Wikoff, H., Rasmussen, A. T.: Effects of heating hypothalamus of dogs by diathermy. J. Neurophysiol. 3, 329–338 (1940).

Hensel, H.: Temperature receptors in the skin. In: J. D. Hardy, A. Ph. Gagge, and J. A. J. Stolwijk, Physiological and Behavioral Temperature Regulation, pp. 442–453. Springfield: Charles C. Thomas Publ. 1970.

Hensel, H.: Neural processes in thermoregulation. Physiol. Rev. 53, 948–1017 (1973).

Herrmann, H. J.: Die Wärmetachypnoe des Meerschweinchens in Abhängigkeit von der Temperatur des vorderen Hypothalamus, des Cervicalmarks, des Lumbalmarks und der Körperoberfläche. Dissertation, Gießen 1972.

Horeyseck, G., Jänig, W.: Activation and inhibition of vasoconstrictors in skin and muscle nerves upon stimulation of skin receptors. Proc. internat. Union physiol. Sci. Vol. IX, p. 260, München 1971.

Hori, T., Nakayama, T.: Effects of biogenic amines on central thermoresponsive neurons in the rabbit. J. Physiol. (Lond.) 232, 71–85 (1973).

Ingram, D. L.: Effects of heating and cooling the hypothalamus on food intake in the pig. Brain Res. 11, 714–716 (1968).

Ingram, D. L., Legge, K. F.: The influence of deep body temperatures and skin temperatures on peripheral blood flow in the pig. J. Physiol. (Lond.) 215, 693–707 (1971).

Ingram, D. L., Legge, K. F.: The influence of deep body temperatures and skin temperatures on respiratory frequency in the pig. J. Physiol. (Lond.) 220, 283–296 (1972a).

Ingram, D. L., Legge, K. F.: The influence of deep body and skin temperatures on thermoregulatory responses to heating of the scrotum in pigs. J. Physiol. (Lond.) 224, 477–487 (1972b).

Iriki, M.: Änderung der Hautdurchblutung bei unnarkotisierten Kaninchen durch isolierte Wärmung des Rückenmarkes. Pflügers Arch. 299, 295–310 (1968).

Iriki, M., Pleschka, K., Walther, O.-E., Simon, E.: Hypoxia and hypercapnia in asphyctic differentiation of regional sympathetic activity in the anesthetized rabbit. Pflügers Arch. 328, 91–102 (1971a).

Iriki, M., Riedel, W., Simon, E.: Regional differentiation of sympathetic activity during hypothalamic heating and cooling in anesthetized rabbits. Pflügers Arch. 328, 320–331 (1971b).

Iriki, M., Riedel, W., Simon, E.: Patterns of differentiation in various sympathetic efferents induced by changes of blood gas composition and by central thermal stimulation in anesthetized rabbits. Jap. J. Physiol. 22, 585–602 (1972).

Iriki, M., Simon, E.: Differential autonomic control of regional circulatory reflexes evoked by thermal stimulation and by hypoxia. Austr. J. Exp. Biol. Med. 51, 283–293 (1973).

Iriki, M., Walther, O.-E., Pleschka, K., Simon, E.: Regional cutaneous and visceral sympathetic activity during asphyxia in the anesthetized rabbit. Pflügers Arch. 322, 167–182 (1971c).

Isenschmid, R., Krehl, L.: Über den Einfluß des Gehirns auf die Wärmeregulation. Naunyn-Schmiedeberg's Arch. 70, 109–134 (1912).

Jacobson, F. H., Squires, R. D.: Thermoregulatory responses of the cat to preoptic and environmental temperature. In: J. D. Hardy, A. Ph. Gagge, and J. A. J. Stolwijk, Physiological and Behavioral Temperature Regulation, pp. 581–596. Springfield: Charles C. Thomas Publ. 1970.

Jessen, C.: Auslösung von Hecheln durch isolierte Wärmung des Rückenmarks am wachen Hund. Pflügers Arch. 297, 53–70 (1967).

Jessen, C.: Spinal and hypothalamic thermodetectors constituting central thermosensitivity in the conscious dog. J. Physiol. (Paris) 63, 306–308 (1971a).

Jessen, C.: Spinal cord and hypothalamus as parallel core sensors of temperature in the dog Int. J. Biometeor. 15, 146–148 (1971b).

Jessen, C., Clough, D. P.: Evaluation of hypothalamic thermosensitivity by feedback signals. Pflügers Arch. 345, 43–59 (1973a).

JESSEN, C., CLOUGH, D. P.: Assessment of spinal temperature sensitivity in conscious goats by feedback signals. J. comp. Physiol. **87**, 75–88 (1973 b).

JESSEN, C., LUDWIG, O.: Spinal cord and hypothalamus as core sensors of temperature in the conscious dog. II. Addition of signals. Pflügers Arch. **324**, 205–216 (1971).

JESSEN, C., MAYER, E. TH.: Spinal cord and hypothalamus as core sensors of temperature in the conscious dog. I. Equivalence of responses. Pflügers Arch. **324**, 189–204 (1971).

JESSEN, C., MCLEAN, J. A., CALVERT, D. T., FINDLAY, J. D.: Balanced and unbalanced temperature signals generated in spinal cord of the ox. Amer. J. Physiol. **222**, 1343–1347 (1972).

JESSEN, C., MEURER, K.-A., SIMON, E.: Steigerung der Hautdurchblutung durch isolierte Wärmung des Rückenmarks am wachen Hund. Pflügers Arch. **297**, 35–52 (1967).

JESSEN, C., SIMON, E.: Spinal cord and hypothalamus as core sensors of temperature in the conscious dog. III. Identity of functions. Pflügers Arch. **324**, 217–226 (1971).

JESSEN, C., SIMON, E., KULLMANN, R.: Interaction of spinal and hypothalamic thermodetectors in body temperature regulation of the conscious dog. Experientia (Basel) **24**, 694–695 (1968).

JONES, R. W.: Biological control mechanisms. In: H. P. SCHWAN, Biological Engineering, Inter-University, Electronic Series, Vol. 9, pp. 87–203. New York: McGraw-Hill Book Comp. 1969.

JUNG, R.: Physiologische Untersuchungen über den Parkinsontremor und andere Zitterformen beim Menschen. Z. ges. Neurol. Psychiat. **173**, 263–332 (1941).

KARIM, F., KIDD, C., MALPUS, C. M., PENNA, P. E.: Left atrial receptors and reflex changes in sympathetic efferent impulse activity. Proc. internat. Union physiol. Sci. Vol. IX, p. 290, München 1971.

KARIM, F., KIDD, C., MALPUS, C. M., PENNA, P. E.: The effects of stimulation of the left atrial receptors on sympathetic efferent nerve activity. J. Physiol. (Lond.) **227**, 243–260 (1972).

KATZ, B., MILEDI, R.: The effect of temperature on the synaptic delay at the neuromuscular junction. J. Physiol. (Lond.) **181**, 656–670 (1965).

KELLER, A. D.: Observations on the localization of the heat regulating mechanisms in the upper medulla and pons. Amer. J. Physiol. **93**, 665 (1930).

KELLER, A. D.: The separation of the heat loss and the heat production mechanisms in chronic preparations. Amer. J. Physiol. **113**, 78–79 (1935).

KELLER, A. D.: Separation in the brain stem of the mechanisms of heat loss from those of heat production. J. Neurophysiol. **1**, 543–557 (1938).

KELLER, A. D.: Temperature regulation disturbances in dogs following hypothalamic ablations. In: C. M. HERZFELD and J. D. HARDY, Temperature, its measurement and control in science and industry. Vol. III, Part III, pp. 571–584. New York: Reinhold Publ. Corporation 1963.

KELLER, A. D., MCCLASKEY, E. B.: Localization, by the brain slicing method, of the level or levels of the cephalic brainstem upon which effective heat dissipation is dependent. Amer. J. Phys. Med. **43**, 181–213 (1964).

KLUGER, M. J., GONZALEZ, R. R., STOLWIJK, J. A. J.: Temperature regulation in the exercising rabbit. Amer. J. Physiol. **224**, 130–135 (1973).

KLUGER, M. J., HEATH, J. E.: Effect of preoptic anterior hypothalamic lesions on thermoregulation in the bat. Amer. J. Physiol. **221**, 144–149 (1971 a).

KLUGER, M. J., HEATH, J. E.: Thermoregulatory responses to preoptic-anterior hypothalamic heating and cooling in the bat, Eptesicus fuscus. J. comp. Physiol. **74**, 340–352 (1971 b).

KLUSSMANN, F. W.: The influence of temperature on the activity of spinal α- and γ-motoneurons. Experientia (Basel) **20**, 450 (1964).

KLUSSMANN, F. W.: Der Einfluß der Temperatur auf die afferente und efferente motorische Innervation des Rückenmarks. I. Temperaturabhängigkeit der afferenten und efferenten Spontantätigkeit. Pflügers Arch. **305**, 295–315 (1969).

KLUSSMANN, F. W., HENATSCH, H.-D.: Der Einfluß der Temperatur auf die afferente und efferente motorische Innervation des Rückenmarks. II. Temperaturabhängigkeit der Muskelspindelfunktion. Pflügers Arch. **305**, 316–339 (1969).

KLUSSMANN, F. W., PIERAU, F.-K.: Extrahypothalamic deep body thermosensitivity. In: J. BLIGH and R. MOORE, Essays on Temperature Regulation, pp. 87–104. Amsterdam: North Holland Publ. Comp. 1972.

KOIZUMI, K., MALCOLM, J. L., BROOKS, C. MCC.: Effect of temperature on facilitation and inhibition of reflex activity. Amer. J. Physiol. **179**, 507–512 (1954).

KOIZUMI, K., USHIYAMA, J., BROOKS, C. McC.: Effect of hypothermia on excitability of spinal neurons. J. Neurophysiol. **23**, 421–431 (1960).

KOSAKA, M., SIMON, E.: Kältetremor wacher, chronisch spinalisierter Kaninchen im Vergleich zum Kältezittern intaker Tiere. Pflügers Arch. **302**, 333–356 (1968a).

KOSAKA, M., SIMON, E.: Der zentralnervöse, spinale Mechanismus des Kältezitterns. Pflügers Arch. **302**, 357–373 (1968b).

KOSAKA, M., SIMON, E., THAUER, R.: Shivering in intact and spinal rabbits during spinal cord cooling. Experientia (Basel) **23**, 385–387 (1967).

KOSAKA, M., SIMON, E., THAUER, R., WALTHER, O.-E.: Effect of thermal stimulation of spinal cord on respiratory and cortical activity. Amer. J. Physiol. **217**, 858–864 (1969a).

KOSAKA, M., SIMON, E., WALTHER, O.-E., THAUER, R.: Response of respiration to selective heating of the spinal cord below partial transection. Experientia (Basel) **25**, 36–37 (1969b).

KULLMANN, R., JUNK, H. G.: Differential sympathetic response during coronary occlusion. Res. exp. Med. **160**, 317–320 (1973).

KULLMANN, R., SCHÖNUNG, W., SIMON, E.: Antagonistic changes of blood flow and sympathetic activity in different vascular beds following central thermal stimulation. I. Blood flow in skin, muscle and intestine during spinal cord heating and cooling in anesthetized dogs. Pflügers Arch. **319**, 146–161 (1970).

LEVY, F.: Recherches expérimentales sur la commande du frisson thermique chez le chien, voies et centres nerveux au niveau de la moelle épinière. Imprimerie Emmanuel Vitte, Lyon (1963).

LIEBERMEISTER, C.: Handbuch der Pathologie und Therapie des Fiebers. Vogel, Leipzig (1875).

LIM, T. P. K.: Central and peripheral control mechanisms of shivering and its effects on respiration. J. Appl. Physiol. **15**, 567–574 (1960).

LIN, M. T., YIN, T. H., CHAI, C. Y.: Effects of heating and cooling of spinal cord on CV and respiratory responses and food and water intake. Amer. J. Physiol. **223**, 626–631 (1972).

LIPTON, J. M.: Effects of preoptic lesions on heat-escape responding and colonic temperature in the rat. Physiol. and Behav. **3**, 165–169 (1968).

LIPTON, J. M.: Behavioral temperature regulation in the rat: effects of thermal stimulation of the medulla. J. Physiol. (Paris) **63**, 325–328 (1971).

LIPTON, J. M.: Thermosensitivity of medulla oblongata in control of body temperature. Amer. J. Physiol. **224**, 890–897 (1973).

MAGOUN, H. W., HARRISON, F., BROBECK, J. R., RANSON, S. W.: Activation of heat loss mechanisms by local heating of the brain. J. Neurophysiol. **1**, 101–114 (1938).

McLEAN, J. A., HALES, J. R. S., JESSEN, C., CALVERT, D. T.: Influences of spinal cord temperature on heat exchange of the ox. Austr. physiol. pharmacol. Soc. **1**, No. 2, 32–33 (1970).

MEURER, K.-A., IRIKI, M., BAUMANN, CH., JESSEN, C.: Kältezittern bei zentraler und peripherer Kühlung. Pflügers Arch. **285**, 63–72 (1965).

MEURER, K.-A., JESSEN, C., IRIKI, M.: Kältezittern während isolierter Kühlung des Rückenmarks nach Durchschneidung der Hinterwurzeln. Pflügers Arch. **293**, 236–255 (1967).

MEYER, H. H.: Theorie des Fiebers und seiner Behandlung. Verh. dtsch. Kongr. inn. Med. Wiesbaden **30**, 15–25 (1913).

MITCHELL, D., ATKINS, A. R., WYNDHAM, C. H.: Mathematical and physical models of thermoregulation. In: J. BLIGH and R. MOORE, Essays on Temperature Regulation, pp. 37–54. Amsterdam: North Holland Publ. Comp. 1972.

MITCHELL, D., SNELLEN, J. W., ATKINS, A. R.: Thermoregulation during fever: change of set-point or change of gain. Pflügers Arch. **321**, 293–302 (1970).

MURAKAMI, N., STOLWIJK, J. A. J., HARDY, J. D.: Responses of preoptic neurons to anesthetics and peripheral stimulation. Amer. J. Physiol. **213**, 1015–1024 (1967).

MURGATROYD, D., HARDY, J. D.: Central and peripheral temperatures in behavioral thermoregulation of the rat. In: J. D. HARDY, A. PH. GAGGE, and J. A. J. STOLWIJK, Physiological and Behavioral Temperature Regulation, pp. 874–891. Springfield: Charles C. Thomas Publ. 1970.

MYERS, R. D.: Hypothalamic mechanisms of pyrogen action in the cat and monkey. In: Pyrogens and fever, pp. 131–153. Edinburgh and London: Churchill Livingstone 1971.

MYERS, R. D., VEALE, W. L.: Body temperature: possible ionic mechanism in the hypothalamus controlling the set point. Science **170**, 95–97 (1970).

MYERS, R. D., YAKSH, T. L.: Thermoregulation around a new "set-point" established in the monkey by altering the ratio of sodium to calcium ions within the hypothalamus. J. Physiol. (Lond.) 218, 609–633 (1971).

MYRHE, K., HAMMEL, H. T.: Behavioral regulation of internal temperature in the lizard Tiliqua scincoides. Amer. J. Physiol. 217, 1490–1495 (1969).

NAKAYAMA, T., EISENMAN, J. S., HARDY, J. D.: Single unit activity of anterior hypothalamus during local heating. Science 134, 560–561 (1961).

NAKAYAMA, T., HAMMEL, H. T., HARDY, J. D., EISENMAN, J. S.: Thermal stimulation of electrical activity of single units of the preoptic region. Amer. J. Physiol. 204, 1122–1126 (1963).

NAKAYAMA, T., HARDY, J. D.: Unit responses in the rabbit's brain stem to changes in brain and cutaneous temperature. J. Appl. Physiol. 27, 848–857 (1969).

NAKAYAMA, T., HORI, T.: Effects of anesthetic and pyrogen on thermally sensitive neurons in the brain stem. J. Appl. Physiol. 34, 351–355 (1973).

NUTIK, ST. L.: Effect of temperature change of the preoptic region and skin on posterior hypothalamic neurons. J. Physiol. (Paris) 63, 368–370 (1971).

NUTIK, ST. L.: Posterior hypothalamic neurons responsive to preoptic region thermal stimulation. J. Neurophysiol. 36, 238–249 (1973a).

NUTIK, ST. L.: Convergence of cutaneous and preoptic region thermal afferents on posterior hypothalamic neurons. J. Neurophysiol. 36, 250–257 (1973b).

OSCARSSON, O.: Functional organization of the spino- and cuneocerebellar tracts. Physiol. Rev. 45, 495–522 (1965).

PIERAU, F.-K.: Zum Mechanismus der Temperaturempfindlichkeit spinaler Neuren der Katze. Habilitationsschrift, Gießen 1971.

PIERAU, F.-K., ALEXANDRIDIS, E., SPAAN, G., OKSCHE, A., KLUSSMANN, F. W.: Der Einfluß von lokalen Temperaturänderungen im pupillomotorischen Kerngebiet der Taube auf die Aktivität der Irismuskulatur. Pflügers Arch. 315, 291–307 (1970a).

PIERAU, F.-K., KLEE, M. R., FABER, D. S., KLUSSMANN, F. W.: Mechanism of cellular thermoreception in mammals. Int. J. Biometeor. 15, 134–140 (1971).

PIERAU, F.-K., KLEE, M. R., KLUSSMANN, F. W.: Effects of local hypo- and hyperthermia on mammalian motoneurones. Fed. Proc. 28, 1006–1010 (1969a).

PIERAU, F.-K., KLUSSMANN, F. W.: Spinal excitation and inhibition during local spinal cooling and warming. J. Physiol. (Paris) 63, 380–382 (1971).

PIERAU, F.-K., SPAAN, G.: Renshaw inhibition during local spinal cord cooling and warming. Experientia (Basel) 26, 978–979 (1970).

PIERAU, F.-K., SPAAN, G., GOLDA, V.: Rekurrente Hemmung und Renshaw-Zellentladung während lokaler Temperaturänderung im Rückenmark der Katze. Pflügers Arch. 316, R 61 (1970b).

PIERAU, F.-K., WURSTER, R. D., SPAAN, G.: Der Einfluß lokaler Temperaturänderungen auf die postsynaptische Hemmung spinaler Moto-Neuren der Katze. Pflügers Arch. 307, R 100 (1969b).

POPOFF, N. F.: Die vegetativen Funktionen des Hundes nach weitgehender Ausschaltung der Einflüsse des Zentralnervensystems. Pflügers Arch. 234, 137–156 (1934).

PRINCE, A. L., HAHN, L. J.: The effect on body temperature induced by thermal stimulation of the heat center in the brain of the cat. Amer. J. Physiol. 46, 412–415 (1918).

RANDALL, W. C., WURSTER, R. D., LEWIN, R. J.: Responses of patients with high spinal transection to high ambient temperatures. J. Appl. Physiol. 21, 985–993 (1966).

RANSON, S. W., MAGOUN, H. W.: The Hypothalamus. Ergebn. Physiol. 41, 56–163 (1939).

RAUTENBERG, W.: Die Bedeutung der peripheren und zentralen Thermosensitivität für die Temperaturregulation der Taube. Habilitationsschrift, Bochum 1967.

RAUTENBERG, W.: Die Bedeutung der zentralnervösen Thermosensitivität für die Temperaturregulation der Taube. Z. Vergl. Physiol. 62, 235–266 (1969).

RAUTENBERG, W.: The influence of skin temperature on the thermoregulatory system of pigeons. J. Physiol. (Paris) 63, 396–398 (1971).

RAUTENBERG, W., NECKER, R., MAY, B.: Thermoregulatory responses of the pigeon to changes of the brain and the spinal cord temperatures. Pflügers Arch. 338, 31–42 (1972).

RAUTENBERG, W., SIMON, E.: Die Beeinflussung des Kältezitterns durch lokale Temperaturänderung im Wirbelkanal. Pflügers Arch. 281, 332–345 (1964).

Rautenberg, W., Simon, E., Thauer, R.: Die Bedeutung der Kerntemperatur für die chemische Temperaturregulation beim Hund in leichter Narkose. I. Isolierte Senkung der Rumpfkerntemperatur. Pflügers Arch. **278**, 337–349 (1963a).

Rautenberg, W., Simon, E., Thauer, R.: Die Bedeutung der Kerntemperatur für die chemische Temperaturregulation beim Hund in leichter Narkose. II. Isolierte Senkung der Hirntemperatur. Pflügers Arch. **278**, 350–360 (1963b).

Rawson, R. O., Quick, K. P.: Evidence of deep-body thermoreceptor response to intra-abdominal heating of the ewe. J. Appl. Physiol. **28**, 813–820 (1970).

Rawson, R. O., Quick, K. P.: Thermoregulatory responses to temperature signals from the abdominal viscera of sheep. J. Physiol. (Paris) **63**, 399–402 (1971a).

Rawson, R. O., Quick, K. P.: Unilateral splanchnotomy: its effect on the response to intra-abdominal heating in the ewe. Pflügers Arch. **330**, 362–365 (1971b).

Rawson, R. O., Quick, K. P.: Localization of the intra-abdominal thermoreceptors in the ewe. J. Physiol. (Lond.) **222**, 665–667 (1972).

Rein, H.: Vasomotorische Regulationen. Ergebn. Physiol. **32**, 28–72 (1931).

Riedel, W., Iriki, M., Simon, E.: Regional differentiation of sympathetic activity during peripheral heating and cooling in anesthetized rabbits. Pflügers Arch. **332**, 239–247 (1972).

Riedel, W., Siaplauras, G., Simon, E.: Intra-abdominal thermosensitivity in the rabbit as compared with spinal thermosensitivity. Pflügers Arch. **340**, 59–70 (1973).

Satinoff, E.: Behavioral thermoregulation in response to local cooling of the rat brain. Amer. J. Physiol. **206**, 1389–1394 (1964).

Satinoff, E., Rutstein, J.: Behavioral thermoregulation in rats with anterior hypothalamic lesions. J. comp. physiol. Psychol. **71**, 77–82 (1970).

Satinoff, E., Shan, S. Y. Y.: Loss of behavioral thermoregulation after lateral hypothalamic lesions in rats. J. comp. physiol. Psychol. **77**, 302–312 (1971).

Schäfer, S. S.: Kennlinien der Meßempfindlichkeiten primärer Muskelspindelafferenzen beim Kältezittern. Pflügers Arch. **309**, 128–144 (1969).

Schäfer, S. S., Schäfer, S.: The behavior of the proprioceptors of the muscle and the innervation of the fusimotor system during cold shivering. Exp. Brain Res. **17**, 364–380 (1973).

Schönung, W., Jessen, C., Wagner, H., Simon, E.: Regional blood flow antagonism induced by thermal stimulation in the conscious dog. Experientia (Basel) **27**, 1291–1292 (1971a).

Schönung, W., Wagner, H., Jessen, C., Simon, E.: Differentiation of cutaneous and intestinal blood flow during hypothalamic heating and cooling in anesthetized dogs. Pflügers Arch. **328**, 145–154 (1971b).

Schwennicke, H.-P.: Die Steuerung der zitterfreien Thermogenese beim Meerschweinchen. Dissertation, Gießen 1972.

Sharp, F. R., Hammel, H. T.: Effects of fever on salivation response in the resting and exercising dog. Amer. J. Physiol. **223**, 77–82 (1972).

Simon, E.: Le rôle de la moelle épinière dans la thermorégulation. Arch. Sci. physiol. (Paris) **21**, 215–233 (1967).

Simon, E.: Spinal mechanisms of temperature regulation. Proc. 24. Internat. Union Physiol. Sci., Vol. VI, pp. 289–290, Washington 1968.

Simon, E.: Regional differentiation of vasomotor activity underlying thermoregulatory adjustments of blood flow. Int. J. Biometeorol. **15**, 219–224 (1971).

Simon, E.: Temperature signals from skin and spinal cord converging on spinothalamic neurons. Pflügers Arch. **337**, 323–332 (1972).

Simon, E., Iriki, M.: Ascending neurons of the spinal cord activated by cold. Experientia (Basel) **26**, 620–622 (1970).

Simon, E., Iriki, M.: Ascending neurons highly sensitive to variations of spinal cord temperature. J. Physiol. (Paris) **63**, 415–417 (1971a).

Simon, E., Iriki, M.: Sensory transmission of spinal heat and cold sensitivity in ascending spinal neurons. Pflügers Arch. **328**, 103–120 (1971b).

Simon, E., Klussmann, F. W., Rautenberg, W., Kosaka, M.: Kältezittern bei narkotisierten spinalen Hunden. Pflügers Arch. **291**, 187–204 (1966).

Simon, E., Kosaka, M., Walther, O.-E.: Afferent influence of changes in spinal cord temperature. Pflügers Arch. **300**, R 43 (1968).

SIMON, E., RAUTENBERG, W., JESSEN, C.: Initiation of shivering in unanesthetized dogs by local cooling within the vertebral canal. Experientia (Basel) 21, 476–477 (1965).

SIMON, E., RAUTENBERG, W., THAUER, R.: Die Bedeutung der Kerntemperatur für die chemische Temperaturregulation beim Hund in leichter Narkose. III. Einfluß der Hirntemperatur auf die Wärmeproduktion bei innerer und äußerer Kältebelastung. Pflügers Arch. 278, 361–373 (1963a).

SIMON, E., RAUTENBERG, W., THAUER, R., IRIKI, M.: Auslösung thermoregulatorischer Reaktionen durch lokale Kühlung im Vertebralkanal. Naturwissenschaften 50, 337 (1963b).

SIMON, E., RAUTENBERG, W., THAUER, R., IRIKI, M.: Die Auslösung von Kältezittern durch lokale Kühlung im Wirbelkanal. Pflügers Arch. 281, 309–331 (1964).

SIMON, E., RAUTENBERG, W., USINGER, W., KOSAKA, M.: Der Einfluß der spinalen Hypothermie auf die spinalen vasomotorischen Zentren. J. Neuro-Viscer. Rel. 31, 350–372 (1970).

SKOGLUND, C. R., UVNÄS, B.: Some inhibitory phenomena in the dorsal root reflex. Acta physiol. scand. 6, 149–159 (1943).

SNELLEN, J. W.: Set point and exercise. In: J. BLIGH and R. MOORE, Essays on Temperature Regulation, pp. 139–148. Amsterdam: North Holland Publ. Comp. 1972.

SPAAN, G., KLUSSMANN, F. W.: Die Frequenz des Kältezitterns bei Tierarten verschiedener Größe. Pflügers Arch. 320, 318–333 (1970).

SPECTOR, N. H., BROBECK, J. R., HAMILTON, C. L.: Feeding and core temperature in albino rats: changes induced by preoptic heating and cooling. Science 161, 286–288 (1968).

SPECTOR, N. H., CORMARÈCHE, M.: Temperature and feeding in dogs: I. Effects of heating and cooling of the vertebral canal upon meal size. J. Physiol. (Paris) 63, 421–424 (1971).

SPERELAKIS, N.: Effects of temperature on membrane potentials of excitable cells. In: J. D. HARDY, A. PH. GAGGE, and J. A. J. STOLWIJK, Physiological and Behavioral Temperature Regulation, pp. 408–441. Springfield: Charles C. Thomas Publ. 1970.

STELTER, W.-J., KLUSSMANN, F. W.: Der Einfluß der Rückenmarkstemperatur auf die Dehnungsantwort tonischer und phasischer α-Motoneurone. Pflügers Arch. 309, 310–327 (1969).

STELTER, W.-J., SPAAN, G., KLUSSMANN, F. W.: Der Einfluß der spinalen und peripheren Temperatur auf die Reflexspannung „roter" und „blasser" Muskeln. Pflügers Arch. 312, 1–17 (1969).

STOLWIJK, J. A. J.: Mathematical model of thermoregulation. In: J. D. HARDY, A. PH. GAGGE, and J. A. J. STOLWIJK, Physiological and Behavioral Temperature Regulation, pp. 703–726. Springfield: Charles C. Thomas Publ. 1970.

STOLWIJK, J. A. J., HARDY, J. D.: Temperature regulation in man—a theoretical study. Pflügers Arch. 291, 129–162 (1966).

STRÖM, G.: Influence of skin temperature on vasodilator response to hypothalamic heating in the cat. Acta physiol. scand. 20, Suppl. 70, pp. 77–81 (1950a).

STRÖM, G.: Effect of hypothalamic cooling on cutaneous blood flow in the unanesthetized dog. Acta physiol. scand. 21, 271–277 (1950b).

STUART, D., ELDRED, E., HEMINGWAY, A., KAWAMURA, Y.: Neural regulation of the rhythm of shivering. In: C. M. HERZFELD and J. D. HARDY, Temperature: Its measurement and control in Science and Industry, Vol. III, Part 3, pp. 545–557. New York: Reinhold Publ. Corporation 1963.

STUART, D., OTT, K., ISHIKAWA, K., ELDRED, E.: The rhythm of shivering. I. General sensory contributions. Amer. J. phys. Med. 45, 61–74 (1966a).

STUART, D., OTT, K., ISHIKAWA, K., ELDRED, E.: The rhythm of shivering. II. Passive proprioceptive contributions. Amer. J. phys. Med. 45, 75–90 (1966b).

STUART, D., OTT, K., ISHIKAWA, K., ELDRED, E.: The rhythm of shivering. III. Central contributions. Amer. J. phys. Med. 45, 91–104 (1966c).

SUDA, I., KOIZUMI, K., BROOKS, C. McC.: Analysis of effects of hypothermia on central nervous system responses. Amer. J. Physiol. 189, 373–380 (1957).

THAUER, R.: Wärmeregulation und Fiebertätigkeit nach operativen Eingriffen am Nervensystem homoiothermer Säugetiere. Pflügers Arch. 236, 102–147 (1935)

THAUER, R.: Der Mechanismus der Wärmeregulation. Ergebn. Physiol. 41, 607–805 (1939).

THAUER, R.: Die Auslösungsmechanismen der chemischen Temperaturregulation. Proc. 22. internat. Union physiol. Sci., Vol. I, Part 1, pp. 390–402 (1962).

THAUER, R.: Der nervöse Mechanismus der chemischen Temperaturregulation des Warmblüters. Naturwissenschaften **51**, 73–80 (1964).

THAUER, R.: Homoiothermie als Fortschritt und Schicksal des Menschen. Jahrbuch d. Max-Planck-Gesellschaft, pp. 39–75, München 1967.

THAUER, R.: Temperature reception in the spinal cord. Symposium on Physiological and Behavioral Temperature Regulation, Abstr. pp. 92–93, New Haven, Conn. 1968.

THAUER, R.: Thermosensitivity of the spinal cord. In: J. D. HARDY, A. PH. GAGGE, and J. A. J. STOLWIJK, Physiological and Behavioral Temperature Regulation, pp. 472–492. Springfield: Charles C. Thomas Publ. 1970.

THAUER, R.: Segmentale Anlage und Cephalisation der nervösen Temperaturregulationsmechanismen. Nova Acta Leopoldina, N.F. **38**, Nr. 211, 251–275 (1973).

THAUER, R., PETERS, G.: Wärmeregulation nach operativer Ausschaltung des „Wärmezentrums". Pflügers Arch. **239**, 483–514 (1938).

THAUER, R., SIMON, E.: Spinal cord and temperature regulation. In: S. ITOH, K. OGATA and H. YOSHIMURA, Advances in climatic physiology, pp. 22–49. Tokyo: Igaku Shoin Ltd. 1972.

TOENNIES, J. F.: Conditioning of afferent impulses by reflex discharges over the dorsal roots. J. Neurophysiol. **2**, 515–525 (1939).

TOTH, D. M.: Temperature regulation and salivation following preoptic lesions in the rat. J. comp. physiol. Psychol. **82**, 480–488 (1973).

WALLIN, B. G., HAGBARTH, K.-E., HONGELL, A., DELIUS, W.: Sympathetic activity in human skin and muscle nerves. Proc. 25. internat. Union physiol. Sci. Vol. IX, p. 594, München 1971.

WALTHER, O.-E., IRIKI, M., SIMON, E.: Antagonistic changes of blood flow and sympathetic activity in different vascular beds following central thermal stimulation. II. Cutaneous and visceral sympathetic activity during spinal cord heating and cooling in anesthetized rabbits and cats. Pflügers Arch. **319**, 162–184 (1970).

WALTHER, O.-E., RIEDEL, W., IRIKI, M., SIMON, E.: Differentiation of sympathetic activity at the spinal level in response to central cold stimulation. Pflügers Arch. **329**, 220–230 (1971a).

WALTHER, O.-E., SIMON, E., JESSEN, C.: Thermoregulatory adjustments of skin blood flow in chronically spinalized dogs. Pflügers Arch. **322**, 323–335 (1971b).

WIT, A., WANG, S. C.: Temperature-sensitive neurons in preoptic anterior hypothalamic region: effects of increasing ambient temperature. Amer. J. Physiol. **215**, 1151–1159 (1968a).

WIT, A., WANG, S. C.: Temperature-sensitive neurons in preoptic anterior hypothalamic region: actions of pyrogen and acetyl salicylate. Amer. J. Physiol. **215**, 1160–1169 (1968b).

WÜNNENBERG, W., BRÜCK, K.: Zur Funktionsweise thermoreceptiver Strukturen im Cervicalmark des Meerschweinchens. Pflügers Arch. **299**, 1–10 (1968a).

WÜNNENBERG, W., BRÜCK, K.: Hyperthermie durch zitterfreie Thermogenese nach Ausschaltung der Regio praeoptica beim Meerschweinchen. Pflügers Arch. **300**, R 45 (1968b).

WÜNNENBERG, W., BRÜCK, K.: Single unit activity evoked by thermal stimulation of the cervical spinal cord in the guinea pig. Nature **218**, 1268–1269 (1968c).

WÜNNENBERG, W., BRÜCK, K.: Studies on the ascending pathways from the thermosensitive region of the spinal cord. Pflügers Arch. **321**, 233–241 (1970).

WÜNNENBERG, W., HARDY, J. D.: Response of single units of the posterior hypothalamus to thermal stimulation. J. Appl. Physiol. **33**, 547–552 (1972).

ZEISBERGER, E., BRÜCK, K.: Central effects of noradrenaline on the control of body temperature in the guinea pig. Pflügers Arch. **322**, 152–166 (1971a).

ZEISBERGER, E., BRÜCK, K.: Comparison of the effects of local hypothalamic acetylcholine and RF-heating on non-shivering thermogenesis in the guinea pig. Int. J. Biometeor. **15**, 305–308 (1971b).

ZEISBERGER, E., BRÜCK, K.: Effects of intrahypothalamically injected noradrenergic and cholinergic agents on thermoregulatory responses. In: E. SCHÖNBAUM and P. LOMAX, The Pharmacology of Thermoregulation, pp. 232–243. Basel: S. Karger 1973.

Rev. Physiol. Biochem. Pharmacol., Vol. 71
© by Springer Verlag 1974

Galactosamine Hepatitis:
Key Role of the Nucleotide Deficiency Period
in the Pathogenesis of Cell Injury and Cell Death

K. DECKER and D. KEPPLER *

Contents

---

* From the Biochemisches Institut an der Medizinischen Fakultät I, Universität Freiburg i. Br.,
D-7800 Freiburg, Germany (FRG). The work of the authors was supported by grants from the
Deutsche Forschungsgemeinschaft, Bonn-Bad Godesberg, Germany.

Introduction

Diseases are the result of injuries to or malfunctioning of certain organs, cells, and ultimately of disturbances of molecular events. It is the rationale and also the justification of the biochemical approach to pathology that diseases originate in the cellular metabolism and are thus amenable to biochemical and biophysical analysis. This should be equally true, whether the interfering events or agents invade the organism from without (bacterial and viral infections, toxic substances, environmental influences) or originate within the body (genetic defects, auto-immune processes, aging). In every case, they affect the organism by interfering with the proper course of enzymic processes and the turnover of cellular structures. For instance, the production of proteins may be inadequate or defective, the activity of enzymes may be changed, or the regulatory mechanism disturbed, resulting in alterations of cellular components or of metabolite fluxes and profiles (Fig. 1).

In genetically determined diseases the primary defect is rather easily detected and defined, since it usually involves deficiency or defective biosynthesis of a single protein species. Nevertheless, it has not so far been possible to elucidate the entire sequence of metabolic events leading from the enzymic defect to the symptoms of the diseases in question. In most other types of diseases, however, the primary lesion is not known and we are faced with a complex of symptoms from which it is extremely difficult to trace the original pathogenetic event. The reason for this difficulty is clear: by the time the macroscopic signs of disease are recognized, the primary lesion may already be obscured by a host of follow-up processes. Although these processes may be typical for a given disease, they involve not only additional metabolic routes but also additional tissues.

Besides the necessity to use laboratory animals for experimentation, it is mainly for the above reason that models of certain diseases have been developed for studies on the biochemistry of pathogenesis, since such models can be deployed at will and studied at any desired stage. It is obvious that most model systems are not likely to reproduce the primary biochemical lesion; at best, they can be expected to effect alterations that are similar or identical to secondary or later

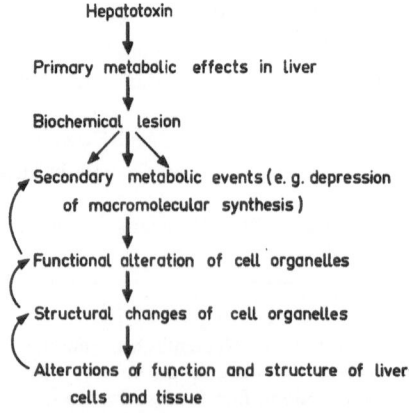

Fig. 1. General sequence of events initiated by a hepatotoxic agent

events in the pathogenetic sequence. The relative position of this point of convergence with the original lesion largely determines the usefulness of the model in advancing understanding of the etiology of a disease. Experimentally induced diseases can, however, be of great value for the elucidation of pathologic processes in more general terms. The use of artificial means to interfere with a metabolic system in a well-defined manner allows one not only to establish the specific consequences of such disturbance but also, and more significantly, to assess the quantitative contribution of an enzymic process within the overall metabolism of the cell.

In view of the coordinated functions of different cell types and organs and of the intracellular compartmentation of the many enzymic reactions, it is essential to an understanding of the metabolic condition of an organism to consider the spatial organization of the processes involved. The activities of the biocatalysts, the enzymes, at any given moment, the turnover of these proteins, and the steady-state concentrations of the metabolites – all these factors determine the rate at which the chemistry of life proceeds. This means that *time* is a crucial factor for the understanding of both the physiologic cell biochemistry and the processes that lead to cell injury and cell death.

The importance of the time factor for the coordinated function of cellular metabolism and its regulation is revealed by quantitative analysis of the rate-controlled parameters. Furthermore, the time course of the changes and distortions introduced by the injurious agent must be studied in order to determine the biochemical events that ultimately cause cell death. We propose that it is not sufficient to define the extent to which activities or concentrations of cellular components are decreased or enhanced; it is also essential to know over what period of time these distortions persist. In order to test this hypothesis it is necessary to determine the duration of a given metabolic aberration *in vivo* and to correlate it with its effects on the function and viability of the target cell or organ. It should then be possible to establish what deviations, both in extent and duration, from the normal state are compatible with the integrity of a given cell.

The liver injury that can be induced in a variety of laboratory animals by the administration of 2-deoxy-2-amino-D-galactose (galactosamine) (KEPPLER et al., 1968) has proved to be very well suited for this kind of study (DECKER and KEPPLER, 1972; DECKER et al., 1973). Galactosamine hepatitis appears to be a unique experimental model, superior in several respects to the many other hepatotoxic agents under study. Within a wide range of galactosamine concentrations, a good dose-effect relationship is obtained if a homogeneous group of animals is used. This also ensures a high degree of reproducibility of both pathologic and biochemical signs of hepatic lesions. Most species respond to galactosamine administration in a very similar manner, although showing different susceptibilities to a given dose (KEPPLER et al., 1968; DECKER et al., 1973). A high degree of sepcificity is observed with regard to the chemical nature of the toxic substance; of the many compounds tested only 2-deoxy-D-galactose was able to induce in guinea pigs liver lesions comparable to those caused by galactosamine (KEPPLER and HÜBNER, 1973). The reason for this specificity is discussed below in more detail.

Conversely, galactosamine affects only the liver; no morphologic or functional alterations have been observed in other cells or organs when doses sufficient

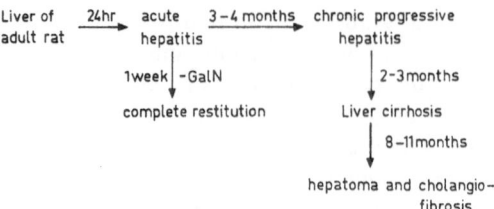

Fig. 2. Time course of the effects of galactosamine (GalN) on rat liver. (From Decker et al., 1973)

to cause liver lesions are given (Keppler et al., 1968). The selectivity of the liver for this amino sugar is a consequence of the specific biochemical composition of the hepatocytes.

Examination of galactosamine-induced liver damage by means of light microscopy and clinical chemistry has revealed features closely resembling those seen in human viral hepatitis. Typical laboratory findings include a rapid loss of liver glycogen, increases in liver-specific enzymes, increased bilirubin in blood plasma, and a reduction in plasma proteins, especially components of the blood-clotting system (Keppler et al., 1968).

The severity and duration of the hepatic lesions elicited by galactosamine (Fig. 2) can be regulated by the amount of amino sugar used, by the schedule of injections, and by several countermeasures, e.g. administration of orotate or uridine (Keppler et al., 1970a; Lesch et al., 1970a, b; Decker et al., 1971). In fact, development of irreversible cell damage can be completely blocked by uridine even at a time when ultrastructural alterations, the release of intracellular proteins, and the loss of glycogen are already in progress (Keppler and Decker, 1971; Decker and Keppler, 1972). It is obvious that these options for regulating the action of galactosamine make it a valuable tool for studying primary biochemical events and for differentiating between the reversible and irreversible alterations observed during the development of this liver injury.

As outlined below in more detail, the biochemical processes involved in the initial stages of this model disease are connected with the metabolism of galactosamine and the biosynthesis of uridylate derivatives. One is dealing, therefore, with an area of cell biochemistry which is well defined and amenable to quantitative analysis. The number of potential secondary metabolic disturbances is restricted to relatively few pathways, so increasing the changes of selecting those molecular mechanisms that are responsible for organelle injuries and ultimately for cell death.

The Biochemical Lesion in Galactosamine-Induced Liver Injury

Metabolism of D-Galactosamine

The first study on the metabolic fate of this amino sugar was conducted by Kuhn et al. (1959) using 1-^{14}C-D-galactosamine. In rats, the intraperitoneally administered galactose analog is taken up predominantly by the liver (Kuhn et al., 1959;

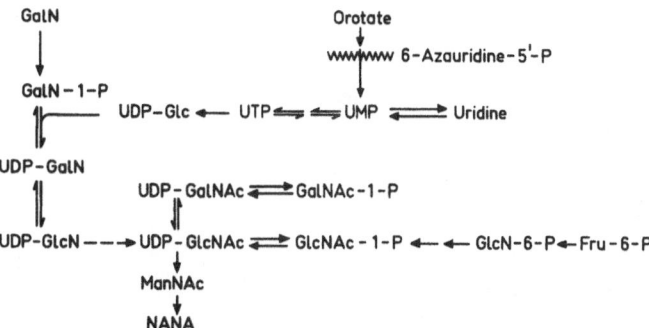

Fig. 3. Metabolism of D-galactosamine (GalN) in liver and its dependence on uridine phosphate biosynthesis. 6-Azauridine-5′-monophosphate is indicated as inhibitor of the synthesis of uridine 5′-monophosphate (UMP) from orotidine monophosphate (CREASEY and HANDSCHUMACHER, 1961). Note the connection of galactosamine metabolism with the physiologic pathways of glucosamine (GlcN) metabolism and N-acetylneuraminic acid (NANA) synthesis. UTP= uridine 5′-triphosphate; UDP-Glc= uridine 5′-diphosphate glucose; GalNAc-1-P= N-acetylgalactos-amine-1-phosphate; Fru-6-P= fructose-6-phosphate; ManNAc= N-acetylmannosamine

WHITE et al., 1965) although significant amounts are eliminated unchanged in the urine (WHITE et. al., 1965). With doses sufficient to produce signs of hepa-titis, about two thirds of the galactosamine recovered from all tissues (KUHN et al., 1959) and about one third of the total dose administered (BAUER et al., 1972) are found in the liver.

Galactosamine is metabolized by enzymes of the galactose pathway (Fig. 3), which are most abundant in liver (CUATRECASAS and SEGAL, 1965; BERTOLI and SEGAL, 1966); the low specificity of galactokinase and UDP-glucose:galactose-1-P uridylyltransferase renders possible the initial steps of galactosamine metabolism (BALLARD, 1966; WALKER and KHAN, 1968). The formation (MALEY et al., 1968) and dose-dependent accumulation (KEPPLER and DECKER, 1969; BAUER et al., 1972) of galactosamine-1-phosphate has been demonstrated in rat liver. The syn-thesis of UDP-galactosamine (MALEY et al., 1968) results from the substitution of galactosamine-1-phosphate for galactose-1-phosphate in the UDP-glucose: galactose-1-phosphate uridylyltransferase reaction (KEPPLER and DECKER, 1969; MALEY, 1970). The low affinity of the enzyme for the nonphysiologic substrate and the limiting amount of UDP-glucose as cosubstrate favor the accumulation of galactosamine-1-phosphate (KEPPLER and DECKER, 1969).

UDP-galactose 4′-epimerase converts UDP-galactosamine to UDP-glucos-amine (MALEY and MALEY, 1959). Neither of these UDP-hexosamines is found in normal liver; both accumulate, however, markedly after galactosamine admini-stration (Fig. 4) (KEPPLER et al., 1970b; BAUER and REUTTER, 1973). UDP-glucos-amine is considered to be a precursor of N-acetylhexosamine monophosphates, UDP-N-acetylgalactosamine, UDP-N-acetylglucosamine and N-acetylneuraminic acid; all these compounds have been identified as metabolites of galactosamine in rat liver (MALEY et al., 1968). The N-acetylated amino sugar nucleotides and phosphates are also present in normal liver; but their levels are greatly increased

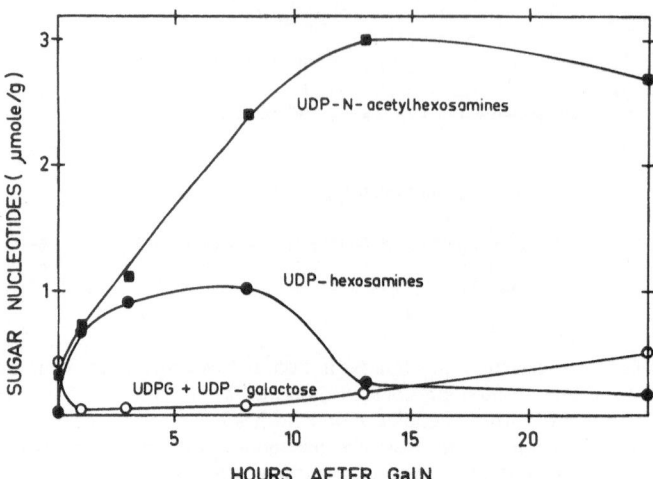

Fig. 4. Galactosamine-induced changes of hepatic uridine diphosphate (UDP) sugars. UDP-N-acetylhexosamines = UDP-GalNAc + UDP-GlcNAc; UDP-hexosamines = UDP-GalN + UDP-GlcN; UDPG = UDP-glucose. (From KEPPLER et al., 1970b)

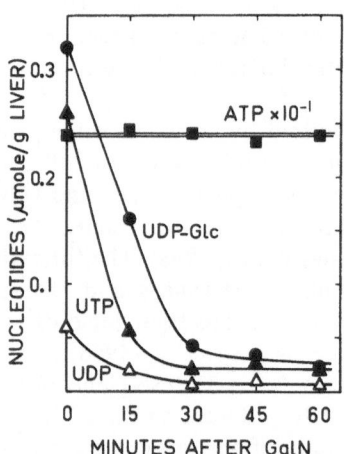

Fig. 5. Rapid depletion of uridine diphosphate glucose (UDP-Glc), uridine 5′-triphosphate (UTP), and uridine 5′-diphosphate (UDP) in rat liver after a dose of galactosamine (1.85 mmol/kg). The adenosine 5′-triphosphate (ATP) content remains in the range of 2.4 μmol/g liver. (From KEPPLER et al., 1974)

following galactosamine administration (KEPPLER et al., 1970b; BAUER et al., 1972). The synthesis and accumulation of UDP-hexosamines and UDP-N-acetylhexosamines is associated with a consumption of the uridylate provided by UDP-glucose and UTP, respectively (Figs. 4, 5).

Alterations in the Contents and Metabolism of Uracil Nucleotides
Caused by Galactosamine

During the initial 30 min after galactosamine administration to rats (1.85 mmol/kg)
UDP-amino sugars increase by 0.85 μmol/g of rat liver while the sum of all acid-
soluble uracil nucleotides rises by only 0.17 μmol/g (KEPPLER et al., 1974) (Fig. 6).
This indicates that the consumption of uridylate greatly exceeds the compensatory
capacity of uridylate *de novo* synthesis (KEPPLER et al., 1970b; DECKER et al.,
1973). The above dose of galactosamine elicits a marked change in the relative
concentrations of the individual uracil nucleotides, leading to a strong depression
of the hepatic contents of UTP, UDP, UMP, UDP-glucose, UDP-galactose, and
UDP-glucuronate (Fig. 5) (KEPPLER et al., 1969; KEPPLER and DECKER, 1969;
KEPPLER et al., 1970a, b, c).

The content of UDP-glucose is depressed from 0.32 to 0.04 μmol/g wet weight
of liver as early as 30 min after galactosamine administration (Fig. 5). UDP-
galactose decreases together with UDP-glucose, and no major deviations from
the UDP-galactose 4'-epimerase equilibrium have been observed (KEPPLER et al.,
1970b). The content of UTP drops to 7% of the control value within 30 min
and remains below 0.03 μmol/g (KEPPLER et al., 1974). The deficiency of UTP
is associated with an equal depression of UDP and of UMP.

Galactosamine-induced UTP deficiency is highly selective; no reduction of
the pools of ATP, GTP (KEPPLER et al., 1974) or CTP (DECKER et al., 1971) has
been observed (Table 1). This is in contrast to many experimental conditions
where a deficiency of UTP is associated with a general depression of nucleoside
triphosphates, e.g. after fructose treatment (BURCH et al., 1969).

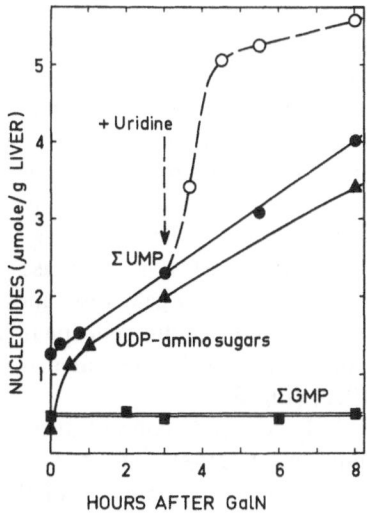

Fig. 6. Content of hepatic uracil and guanine nucleotides after galactosamine administration.
ΣUMP= sum of all acid-soluble uracil 5'-nucleotides; ΣGMP= sum of all acid-soluble guanine
5'-nucleotides; UDP-amino sugars= UDP-N-acetylhexosamines+ UDP-hexosamines. (From
KEPPLER et al., 1974)

Table 1. Selectivity with regard to other ribonucleoside triphosphates of galactosamine-induced uridine triphosphate deficiency. Liver nucleotides measured in rats 3 h after injection of galactosamine (1.85 mmol/kg). (From KEPPLER et al., 1974)

Nucleotide measured	Control	Galactosamine
UTP	0.26 ± 0.04	0.02 ± 0.005
CTP	0.08 ± 0.01	0.17 ± 0.02
ATP	2.44 ± 0.11	2.24 ± 0.13
GTP	0.31 ± 0.03	0.37 ± 0.07

The total content of uracil nucleotides ($\sum$UMP) in liver, which is determined as 5'-UMP after enzymic hydrolysis of allacid-soluble nucleotides, starts to increase as soon as uridine phosphate levels drop to less than about 30% of those in the controls (KEPPLER et al., 1970b). This increase of $\sum$UMP results from a relief of feedback inhibition of *de novo* pyrimidine biosynthesis under conditions of UTP deficiency (KEPPLER et al., 1970b; DECKER et al., 1973). The elevated rate of *de novo* pyrimidine biosynthesis can be suppressed by inhibitors of this pathway, particularly 6-azauridine (KEPPLER et al., 1970b). The available evidence suggests that glutamine-dependent carbamylphosphate synthetase is the site of feedback control of *de novo* pyrimidine biosynthesis by UTP (LEVINE et al., 1971; SMITH et al., 1973). Galactosamine induced hepatic UTP deficiency serves as a tool for studies on the regulation of this pathway.

Effects of 2-Deoxy-D-Galactose, D-Glucosamine, and D-Galactose

Formation and accumulation of the respective UDP-sugar derivatives, causing a depression of UTP, UDP, and UMP, has also been found after administration of 2-deoxy-D-galactose, D-glucosamine and, to some extent, D-galactose. The metabolism of *2-deoxy-D-galactose* on the galactose pathway (Fig. 7) was first demonstrated in yeast and higher plants by FISCHER and WEIDEMANN (1964a, b). The metabolic conversion of 2-deoxy-D-galactose in liver is again due to the low specificity of galactokinase (BALLARD, 1966; WALKER and KHAN, 1968) and of UDP-glucose:galactose-1-phosphate uridylyltransferase (KEPPLER et al., 1970b).

The accumulation of UDP-2-deoxy-D-galactose and UDP-2-deoxy-D-glucose functions as a trapping mechanism for uridine phosphate (KEPPLER et al., 1970b); the resulting UTP deficiency has been demonstrated in rat and guinea-pig liver (KEPPLER and HÜBNER, 1973) (Fig. 8).

D-Glucosamine, particularly at high concentrations, causes an increase of UDP-N-acetyl-D-glucosamine and UDP-N-acetyl-D-galactosamine in liver (KORNFELD et al., 1964; BATES et al., 1966). A reduction of the uridine phosphate and UDP-hexose pools occurs as a consequence of D-glucosamine metabolism (Fig. 9) in both liver (KEPPLER et al,. 1970b) and tumor cells (BEKESI and WINZLER, 1969). The extent of the glucosamine-induced uridine phosphate trapping is strongly dependent on the enzyme patterns of the respective tissues. The effect of glucos-

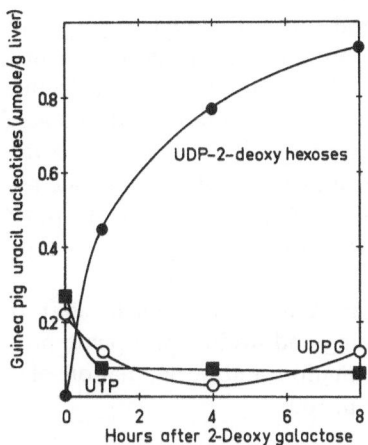

Fig. 7. Pathway of 2-deoxy-D-galactose metabolism in liver. (From KEPPLER and HÜBNER, 1973)

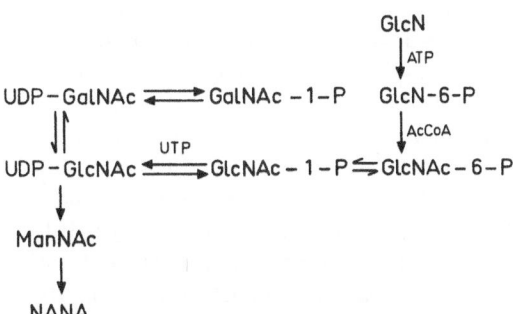

Fig. 8. Uridine triphosphate (UTP) and uridine diphosphate glucose (UDPG) deficiency as a consequence of uridine diphosphate (UDP) 2-deoxy-D-hexose accumulation. A single dose of 2-deoxy-D-galactose (3.5 mmol/kg) was administered to guinea pigs. (From KEPPLER and HÜB-NER, 1973)

$$
\begin{array}{ccc}
 & & GlcN \\
 & & \downarrow {\small ATP} \\
UDP-GalNAc \rightleftharpoons GalNAc-1-P & GlcN-6-P \\
\big\Downarrow\big\Uparrow \qquad\quad {\small UTP} & & \downarrow {\small AcCoA} \\
UDP-GlcNAc \rightleftharpoons GlcNAc-1-P \rightleftharpoons GlcNAc-6-P \\
\downarrow \\
ManNAc \\
\downarrow \\
NANA
\end{array}
$$

Fig. 9. Metabolism of exogenous glucosamine (GlcN). The N-acetylation of glucosamine-6-phosphate requires acetyl coenzyme A (AcCoA). The formation of uridine diphosphate N-acetyl-glucosamine (UDP-GlcNAc) consumes uridine triphosphate (UTP); further abbreviations are explained in the legend to Fig. 3

amine on liver is much less pronounced than that of the galactose analogs, while in fibroblasts only glucosamine is an efficient uridylate trapping agent (SCHOL-TISSEK, 1971).

Besides the non-physiologic sugar analogs, including 2-deoxy-D-galactose and the non-acetylated hexosamines, D-*galactose* can also lead to lowered hepatic uridine phosphate and UDP-glucose pools. An extensive accumulation of UDP-galactose in liver, however, is seen only when UDP-galactose 4'-epimerase is made the limiting enzyme of hepatic galactose metabolism (KEPPLER et al., 1970d).

The Mechanism of Uridylate Trapping

Exogenous supply of several sugars that can be phosphorylated at carbon 1 and subsequently uridylylated produces a change in the distribution of liver uracil nucleotides. The trapping of the uridylate moiety from UTP and UDP-glucose is associated with the following major interrelated changes (KEPPLER et al., 1970b): (a) accumulation of the respective UDP-sugar derivatives, (b) a marked decrease of UTP, UDP and UMP, and (c) a subsequent increase in the sum of hepatic uracil nucleotides. Depletion of uridine phosphate pools is followed by UDP-glucose and UDP-galactose deficiency; D-galactose, of course, increases the UDP-galactose pool.

Sugar analogs or sugars that trap uridylate shift the equilibrium between uridine phosphate-consuming and uridine phosphate-producing reactions toward consumption by rapid formation and accumulation of the UDP derivatives of the sugar in question (see p. 100).

Modification of Uridylate Trapping by Exogenous Pyrimidine
Nucleotide Precursors

A very efficient way to counteract the trapping of uridylate is to administer pyrimidine nucleotide precursors. The initial rate of galactosamine-induced uridine phosphate trapping in liver is approximately 1.7 µmol/g/h (KEPPLER et al., 1974). Pyrimidine nucleotide precursors will partially or completely compensate for this inadequate capacity, depending on the rate at which they increase the uracil nucleotide pool. Uridine is the only precursor that is taken up and phosphorylated in liver at a rate sufficient to reverse or fully compensate a galactosamine-induced uridine phosphate deficiency (KEPPLER et al., 1974). Uridine increases the total uracil nucleotides about 2.5 times faster than orotate, but the effect of orotate is more prolonged. Orotate pretreatment has been successfully used to counteract induction of uridine phosphate deficiency and to prevent the liver injury induced by galactosamine (KEPPLER et al., 1970a; DECKER et al., 1971). Ureidosuccinate and carbamyl phosphate injected intraperitoneally are much less efficient than uridine and orotate (KEPPLER et al., 1974); no significant effect on uracil nucleotide levels has been detected after uracil administration.

Genetically Determined UDP-Hexose Deficiency in Hepatoma Cells

In contrast to the depletion of uridine phosphates, UDP-glucose and UDP-galactose induced by sugar analogs, a permanent deficiency of UDP-glucose and UDP-galactose has been observed recently in two rat ascites hepatoma lines originally derived from liver (KEPPLER and SMITH, 1974). The UTP content of these cells is 2.6 times higher than in liver but the contents of UDP-glucose and UDP-galactose are approximately 15% of those measured in liver. The selectivity of this permanent UDP-hexose deficiency is established by the fact that UDP-N-acetylhexosamines and UDP-glucuronate are in the same range as in rat liver (KEPPLER and SMITH, 1974). The loss of glycogen in both cell lines appears to be one of the consequences of UDP-glucose deficiency. The genetic defect in the nucleotide pattern of these fast growing hepatoma cells is important for an understanding of the consequences of galactosamine-induced hepatic UDP-hexose deficiency.

Inhibition of Macromolecular Syntheses as a Consequence of UTP and UDP-Hexose Deficiency

Ribonucleic Acids

The template activity of chromatin, the activity of partially purified RNA polymerase, and the ability of isolated nuclei to incorporate labeled ribonucleoside triphosphates into RNA *in vitro* have been found to be unchanged in preparations from rat liver 2 h after galactosamine administration *in vivo* (SHINOZUKA et al., 1973a). The total amount of hepatic RNA, and particularly of UMP from RNA, is not altered significantly (KEPPLER et al., 1970b). An effect of galactosamine metabolites has been ruled out by uridine reversal of both UTP deficiency and inhibition of RNA synthesis in the presence of high concentration of amino sugar metabolites (DECKER et al., 1973; REYNOLDS and REUTTER, 1973; KEPPLER et al., 1974). The measurement of UTP separate from other uridine phosphates has shown a depression of the hepatic content to 0.02 µmol/g. This corresponds to an estimated intracellular concentration of 0.027 mM, depressed from a normal value of about 0.35 mM (KEPPLER et al., 1974). CHAMBON (1972) has determined a K_m value of about 0.05 mM for UTP with both types of rat liver nuclear RNA polymerases. This means that these enzymes are saturated with respect to UTP under physiological conditions but that after galactosamine treatment only about one fourth of the maximal rate of the polymerases can be expected. It has been established that the rat liver nuclear membrane is freely permeable to ribonucleoside triphosphates (SIEBERT, 1972). Inhibition of RNA synthesis following the administration of galactosamine would therefore seem a logical consequence. However, until recently definite evidence of this relationship was lacking.

Labeled guanosine has been used as the precursor for *in vivo* measurements of RNA synthesis since neither the content of GTP nor that of total guanine nucleotides is changed significantly by galactosamine (Table 1, Fig. 6) (KEPPLER et al., 1974). By contrast, labeled uridine or orotate, as routinely used for the measurement of RNA synthesis, would be incorporated into RNA of galactosamine-treated liver from a strongly reduced UTP pool, thus giving higher specific

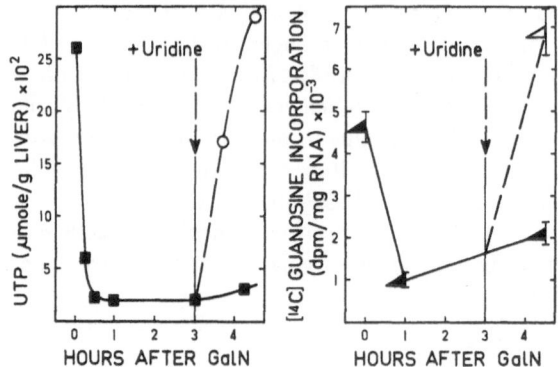

Fig. 10. Changes in the uridine triphosphate (UTP) content of rat liver (left) associated with a reversible depression of ribonucleic acid (RNA) synthesis (right). (From KEPPLER et al., 1974)

UTP radioactivity than controls. Furthermore, uridine uptake and UTP-consuming processes other than RNA synthesis, particularly the formation of UDP-amino sugars, differ greatly under the conditions used. Incorporation of uridine and orotate into RNA would therefore lead to erroneous measurements in galactosamine-treated animals.

RNA synthesis, measured during a 30-min period by incorporation of ^{14}C-guanosine into isolated liver RNA, is depressed to 21 % of the controls when the UTP content is reduced to 0.02 µmol/g of liver (KEPPLER and DECKER, 1972; DECKER et al., 1973; KEPPLER et al., 1974). This inhibition of RNA synthesis is promptly reversed by uridine administration and is paralleled by changes in the UTP content (Fig. 10).

UTP deficiency for relatively short periods of time will affect the various RNA species differently, depending on their respective turnover rates. Inhibition of the dexamethasone-induced increase of tyrosine aminotransferase by galactosamine suggests that the synthesis of messenger RNA is greatly depressed (REYNOLDS and REUTTER, 1973). SHINOZUKA et al. (1973b) have presented evidence that indicates inhibition of ribosomal RNA synthesis. Some disaggregation and loss of polyribosomes has been demonstrated as early as 4 h after galactosamine treatment (SHINOZUKA et al., 1973a).

Proteins

Inhibition of protein synthesis as a result of galactosamine action on liver is indicated by the decreased incorporation of labeled amino acids into liver proteins (REUTTER et al., 1969; MONIER and WAGLE, 1971; KOFF et al., 1971; ANUKARA-HANONTA et al., 1973). ANUKARAHANONTA, SHINOZUKA and FARBER (1973) have shown that protein synthesis, measured over very short time periods, is depressed to about 50 % of normal as early as 30 min after galactosamine administration to rats at 0.93 mmol/kg. At this dose, the minimum incorporation (10% of controls) is reached after 2 h and sustained at this level for 3 to 6 h. In accord with these findings is the reduced synthesis and concentration of plasma proteins, particularly coagulation factors, produced by the liver (KEPPLER et al., 1968; REUTTER et al., 1969; GRÜN et al., 1972; MÜLLER-BERGHAUS and REUTER, 1973). It appears most significant that uridine prevents or almost completely reverses the inhibition of protein synthesis caused by galactosamine (ANUKARAHANONTA et al., 1973). This indicates that the depression of protein synthesis is not related to the presence of galactosamine metabolites but is clearly a consequence of the deficiency of UTP, UDP-glucose, and UDP-galactose. What is not clear at present is whether this deficiency interferes with protein synthesis directly (ANU-KARAHANONTA et al., 1973) or whether the effect is mediated via inhibition of RNA synthesis. As far as glycoproteins are concerned, a defect in glycosylations might be an additional mechanism contributing to reduced synthesis.

Carbohydrate-Containing Macromolecules

For the synthesis of glycogen, glycoproteins, and glycolipids, as well as of glyco-lipoproteins, UDP sugars are required as substrates. Synthesis of the carbohydrate moieties of these macromolecules could be affected by galactosamine administration for two reasons: (a) The greatly reduced concentrations of UDP-glucose, UDP-galactose and, to a minor extent, of UDP-glucuronate will slow down glycosylations, depending on the K_m value of the relevant transferase; (b) the disturbed pattern of sugar nucleotides includes an increase in UDP-glucosamine and UDP-galactosamine at a time when UDP-glucose and UDP-galactose levels are low. This disturbance may be associated with an incorporation of non-acetylated amino sugars, as discussed earlier (KEPPLER et al., 1970b, c; MALEY, 1970). However, the pathogenetic significance of such incorporated amino sugars has not been established under *in vivo* conditions. However, in view of the following facts: (a) the instant reversal of the pathogenetic process by uridine and its prevention by orotate, (b) the induction of liver injury in rats by the combined action of 6-azauridine and 2-deoxy-D-galactose (DECKER et al., 1971), (c) the imitation of galactosamine-induced injury by 2-deoxy-D-galactose in the guinea pig (KEPPLER and HÜBNER, 1973), it is doubtful whether the unnaturally incorporated amino sugars will have a pathological effect.

The available evidence strongly suggests that the lowered hepatic glycogen content after galactosamine administration (Fig. 11) (KEPPLER et al., 1968) is a consequence of the low concentration of UDP-glucose available as substrate for glycogen synthase (KEPPLER and DECKER, 1969). Orotate pretreatment has been shown to strongly counteract both depletion of the UDP-glucose pool and depression of the glycogen content (KEPPLER et al., 1970a; DECKER et al., 1971). Inhibition of glycoprotein synthesis is indicated by the markedly reduced incorporation of ^{14}C-mannose and ^{14}C-galactose into glycoproteins of liver and serum (REUTTER et al., 1969). It is difficult, however, to differentiate between inhibition of synthesis of the polypeptide and of the oligosaccharide moiety.

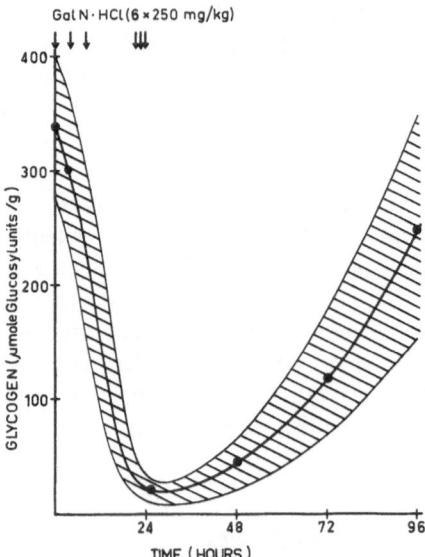

Fig. 11. Time course of the hepatic glycogen content after galactosamine administration to rats. Six doses of 1.16 mmoles (= 250 mg) each of galactosamine hydrochloride were injected at the times indicated by vertical arrows (KEPPLER et al., 1968). The hatched area indicates the standard deviation. (From LESCH et al., 1970b)

Altered Plasma Enzyme Activities and Morphologic Signs of Liver Injury Induced by D-Galactosamine and 2-Deoxy-D-Galactose

Plasma Enzyme Activities, Bilirubin, Coagulation Factors

Several parameters in the blood plasma can be used to quantitate hepatocellular injury and necrosis. The release of intracellular macromolecules into the plasma is commonly determined by measurement of enzyme activities. The liver specificity of these assays depends on the distribution of the enzyme in question among the various tissues. The activity measured in serum or plasma depends not only on the amount of enzyme leaking out from the injured or necrotic cell but also on the rates at which it is degraded and eliminated from the serum. Sorbitol dehydrogenase (L-iditol dehydrogenase) and glutamate dehydrogenase are considered to indicate liver damage rather specifically (GERLACH and HIBY, 1970; SCHMIDT, 1970); however, alanine and aspartate transaminases (DE RITIS et al., 1956) and various other enzymes are also widely used (MÖHR et al., 1971).

A significant increase to about twice the control activity of sorbitol dehydrogenase and of transaminases is already observed in rats 3 h after administration of a single dose of more than 1 mmol/kg of galactosamine (LESCH et al., 1970b; REUTTER and BACHMANN, 1971). Plasma enzyme activities reach a peak between 24 and 48 h (LESCH et al., 1970b) (Table 2). The combined administration of

Table 2. Prevention of galactosamine-induced liver injury by uridine. Plasma enzyme activities (mU/ml) and bilirubin (mg/100 ml) were determined 24 h after a single dose of galactosamine given to rats (1.85 mmol/kg). Uridine (4 mmol/kg) was given 3, 5, and 8 h after galactosamine. (From KEPPLER, 1973)

	Control	Galactosamine + Uridine	Galactosamine
Sorbitol dehydrogenase	3 ± 2	4 ± 1	139 ± 56
Glutamate dehydrogenase	3 ± 1	5 ± 1	129 ± 48
Alanine transaminase	42 ± 8	45 ± 9	720 ± 304
Aspartate transaminase	38 ± 5	75 ± 10	659 ± 279
Total bilirubin	< 0.2	0.1	1.2 ± 0.4
Direct reacting bilirubin (%)	—	—	71 ± 9%

Table 3. Liver injury induced by repeated doses of 6-azauridine + 2-deoxy-D-galactose. Enzyme activities (mU/ml) and bilirubin (mg/100 ml) were measured in rat plasma 24 h after the initial dose. Doses of 6-azauridine (1.1 mmol/kg), 2-deoxy-D-galactose (3.5 mmol/kg), D-glucosamine (3.5 mmol/kg), and uridine (3 mmol/kg) were administered 1, 2, 3 or 4 times as indicated at 0, 2, 4, and 10 h, respectively. (From DECKER et al., 1971)

	Sorbitol dehydrogenase	Alanine transaminase	Total bilirubin
Control	3 ± 2	42 ± 8	< 0.20
6-Azauridine + 2-deoxygalactose			
1 ×	3 ± 1	54 ± 1	0.15 ± 0.08
2 ×	21 ± 2	92 ± 6	0.15 ± 0.07
3 ×	51 ± 32	202 ± 71	0.44 ± 0.17
4 ×	175 ± 97	451 ± 220	2.17 ± 1.39
6-Azauridine + 2-deoxygalactose + D-glucosamine			
4 ×	900 ± 432	960 ± 267	15.4 ± 3.3
6-Azauridine + 2-deoxygalactose + uridine			
3 ×	3 ± 1	40 ± 4	—

6-azauridine and 2-deoxy-D-galactose also produces a dose-dependent increase of enzymes in rat plasma (Table 3) (DECKER et al., 1971). Only 4 h after injection of 2-deoxy-D-galactose alone into guinea pigs, a 3-fold and 5-fold increase in sorbitol and glutamate dehydrogenase, respectively, has been measured (KEPPLER and HÜBNER, 1973). These enzyme activities are greatly increased in guinea-pig plasma after 24 h (Table 4).

An elevated concentration of plasma bilirubin has been found 24 and 48 h after administration of galactosamine (Table 2) (KEPPLER et al., 1968; DECKER et al., 1971). Approximately 73% of total bilirubin is direct-reacting and therefore conjugated (DECKER and KEPPLER, 1972; KEPPLER and HÜBNER, 1973). The sugar

Table 4. Liver injury induced in guinea pigs by 2-deoxy-D-galactose. Plasma enzyme activities (mU/ml) and bilirubin (mg/100 ml) were determined 24 h after the initial dose. 2-Deoxy-D-galactose (3.5 mmol/kg) was administered at 0, 2, and 6 h. (From Keppler and Hübner, 1973)

	Control	2-Deoxy-D-galactose
Sorbitol dehydrogenase	49 ± 14	793 ± 280
Glutamate dehydrogenase	4 ± 2	135 ± 41
Aspartate transaminase	38 ± 13	325 ± 117
Total bilirubin	< 0.1	2.2 ± 0.8
Direct reacting (%)	—	73 ± 10

moieties of the bilirubin conjugates have not been analyzed but the high percentage of conjugates does not suggest that the lowered UDP-glucuronate content plays a significant role in galactosamine-treated liver (Keppler et al., 1970b). In liver slices, however, strong inhibition of o-aminophenol glucuronide synthesis in the presence of galactosamine has been demonstrated (Hänninen and Marniemi, 1970; Marniemi et al., 1973).

Many of the coagulation factors in serum have a short half-life and are synthesized by the liver. A decrease in the activity and amount of coagulation factors can therefore be interpreted as a sign of liver injury (Colombi et al., 1968). The markedly prolonged prothrombin time as well as the hemorrhages observed indicate impaired function of galactosamine-treated liver (Keppler et al., 1968). Detailed studies of the disturbances in blood coagulation associated with galactosamine-induced liver injury in rats (Grün et al., 1972) and rabbits (Müller-Berghaus and Reuter, 1972) indicate a strongly decreased synthesis of coagulation factors, associated with additional consumption coagulopathy.

Light-Microscopic and Ultrastructural Changes in Liver

A single dose of galactosamine (1 mmol/kg) induces after 4 h light-microscopically detectable alterations characterized by acidophilic degeneration of the cytoplasm. Necrosis of single hepatocytes is evident after 6 h (Koff et al., 1971). After 24–26 h all signs of a spotty necrotic hepatitis are found (Keppler et al., 1968; Medline et al., 1970). Foci of hepatocellular necrosis are disseminated over the lobules and the hepatocytes are replaced by inflammatory infiltrates consisting mainly of segmented leukocytes, lymphocytes, and plasma cells. Acidophilic degeneration and acidophilic bodies are frequently observed. Kupffer cells are enlarged and appear more numerous. Inflammatory infiltrations are pronounced in periportal areas. Bile ductules in and around portal tracts are increased in number. In surviving animals normal histologic features of the liver are observed within 7 to 12 days (Lesch et al., 1970b).

Repeated injections of galactosamine over a period of 6 months induce liver cirrhosis in the rat (Lesch et al., 1970a). The lobular architecture is progressively changing and is characterized by an increase of fibrous connective tissue surrounding areas of hyperplasia and regeneration. Cholangiofibrosis and hepatomas

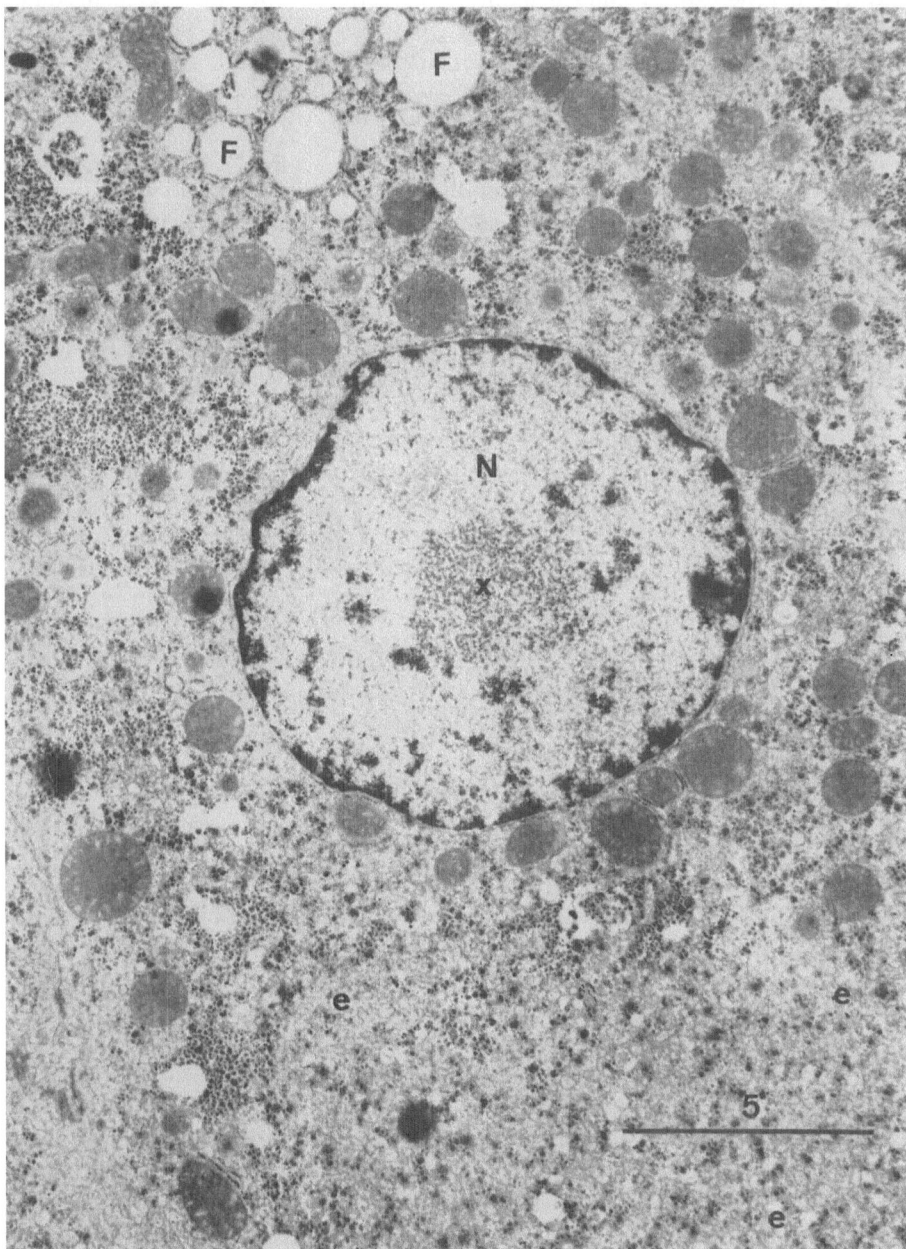

Fig. 12. Guinea-pig liver 10 h after the initial dose of 2-deoxy-D-galactose. The galactose analog (3.5 mmol/kg) was injected at 0, 2, and 6 h. Fat droplets (F); increased smooth endoplasmic reticulum (e); fragmentation of the nucleolus (N) (×); (× 7000). (From KEPPLER and HÜBNER, 1973)

have been observed 13 and 18 months, respectively, after daily galactosamine administrations (LESCH et al., 1973).

Ultrastructural changes have been observed as early as 30 min after galactosamine administration (SHINOZUKA et al., 1973b). The onset of fragmentation of nucleoli and the disruption of nucleolonema (SHINOZUKA et al., 1973a, b) coincides with the rapid induction of UTP deficiency and inhibition of RNA synthesis (Fig. 10). After 2 h nucleolar fragmentation is more marked; breakdown of polyribosomes is pronounced 4 h after a single galactosamine dose (SHINOZUKA et al., 1973a). An increasing number of autophagic vacuoles containing mitochondria, glycogen, and fibrillar and granular material is observed in many hepatocytes after 6 h (KOFF et al., 1971). Normal hepatocytes are no longer seen 24 h after initiation of a series of galactosamine injections (MEDLINE et al., 1970). Many cells appear ballooned, and glycogen rosettes have disappeared. The rough endoplasmic reticulum is dilated and diminished in amount, and the smooth reticulum shows hypertrophy (MEDLINE et al., 1970; SCHARNBECK et al., 1972). All of these ultrastructural changes in hepatocytes have also been described after 2-deoxy-D-galactose administration to guinea pigs (Fig. 12) (KEPPLER and HÜBNER, 1973).

The Impact of the Time Factor on Induction, Prevention and Reversal of Cell Damage

Critical Ranges of Uracil Nucleotide Concentrations and Critical Time Periods

UTP, UDP-glucose, and UDP-galactose are indispensable substrates in the synthesis of cellular components such as nucleic acids, glycoproteins, and glycolipids (Fig. 13). The complete loss of any one of these biosynthetic intermediates has to be considered incompatible with the integrity of structure and function of a viable liver cell, unless the turnover rate of the cellular components is approaching zero. It is consistent, therefore, that mutations in which any one of these nucleotides is completely absent have never been found. A strongly depressed level of UDP-glucose and UDP-galactose, however, appears to be compatible with the viability and rapid growth of certain cell lines originally derived from liver (KEPPLER and SMITH, 1974). The content of UDP-galactose in AS-30D and Novikoff ascites hepatoma cells is 0.017 and 0.016 µmol/g wet weight, respectively, as compared to 0.090 µmol/g in liver. The corresponding figures for UDP-glucose are 0.052 and 0.041 versus 0.320 µmol/g (KEPPLER and SMITH, 1974). UDP-hexose levels in a similar range as in these hepatoma lines have been determined in brain and skeletal muscle (KEPPLER et al., 1970e). Where UDP-glucose and UDP-galactose deficiencies are induced in liver by galactosamine and 2-deoxy-D-galactose, UDP-hexose levels are in the same range as in the hepatoma cell lines (KEPPLER et al., 1970b; KEPPLER and HÜBNER, 1973). The data indicate that depression of UDP-hexose levels to about 15% of the normal hepatic value is still compatible with cell viability, and even with cell growth, at least in the two

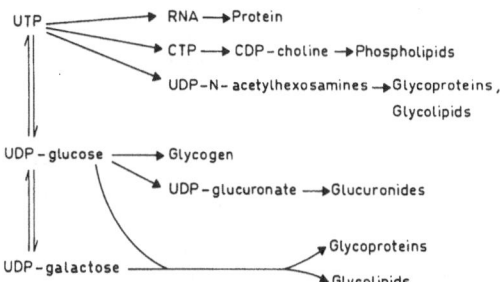

Fig. 13. Biosyntheses dependent on uridine triphosphate (UTP), uridine diphosphate (UDP) glucose, and UDP-galactose as substrates. CTP= cytidine 5′-triphosphate. (From DECKER et al., 1973)

ascites hepatoma lines. However, a UDP-glucose level of less than 0.060 μmol/g appears to be below the concentration required for net synthesis and storage of glycogen. This is consistent with the range of K_m values of hepatic glycogen synthase a for UDP-glucose (MERSMANN and SEGAL, 1967; DE WULF et al., 1970).

There are well-known examples, e.g. ischemia and oxygen deficiency, demonstrating that not only threshold concentrations but also critical time periods determine whether cell damage in brain and other tissues is reversible or not (THORN, 1966). Threshold concentrations and critical time periods can be established only in defined systems under reproducible conditions. For instance, the critical period for UTP deficiency for normal liver of female young adult rats will differ from that for regenerating liver (see p. 99); so that a *deficit period* established for liver may not be applicable to brain or even to hepatoma. Furthermore, the criteria used for the quantitative detection of cell injury (e.g. activities of hepatic enzymes in plasma at a certain time, or quantifiable morphologic changes) must be the same when critical deficit periods are studied under varying experimental conditions. Enzyme activities in plasma, measured when galactosamine- or 2-deoxy-D-galactose-induced liver-cell necrosis is evident (e.g. after 24 h), have been used to assess the effect of agents which prevent or modify the period of uridine phosphate deficiency (KEPPLER et al., 1970a; DECKER et al., 1971) (Table 2). Various observations on the experimental liver injury caused by uridine phosphate deficiency have shown that UTP levels down to 30% of normal do not cause irreversible cell damage or cell death (DECKER et al., 1971); below this range of the hepatic UTP content, corresponding to about 0.080 μmol/g, UTP-dependent processes are progressively slowed down until they are incompatible with cell viability when they persist over several hours. Products of UTP-dependent biosyntheses will be more readily affected the higher the turnover of the product and the lower the affinity of the relevant enzyme for UTP. The duration of severe UTP deficiency that will result in irreversible cell injury and cell death is more than 2 h (DECKER and KEPPLER, 1972) and less than 5 h (FARBER et al., 1973). That hepatic UTP deficiency can be tolerated for as long as 3 h is indicated by light microscopy and plasma enzyme activities at 12 h (FARBER et al., 1973) and 24 h (Table 2). Although ultrastructural lesions and a slight elevation of plasma enzyme activities have been detected 3 h after galactosamine administration, these changes are evidently readily reversible.

Uridine Reversal of UTP and UDP-Hexose Deficiency:
Effect on Macromolecular Syntheses and Liver Injury

Galactosamine-induced deficiency of UTP and UDP-hexoses can be reversed rapidly and at any time by administration of uridine (KEPPLER and DECKER, 1971). Hence it is possible to inhibit UTP-dependent processes *in vivo* for periods corresponding to the time interval between galactosamine and uridine injection (KEPPLER et al., 1974) (Fig. 10). This enables the consequences of different periods of UTP deficiency in liver to be studied. The deficit period required to cause irreversible cell injury has been analyzed in this way (DECKER et al., 1973; FARBER et al., 1973) (Table 2). Uridine reversal of UTP and UDP-hexose deficiency 2 h and even 3 h after galactosamine administration prevents the development of hepatocellular necrosis. None of the other protective agents, including galactose (REUTTER et al., 1973), is capable of preventing irreversible liver cell injury after a deficit period of 2 h.

Uridine has been shown to reverse completely the inhibition of hepatic protein synthesis 2 h (ANUKARAHANONTA et al., 1973) and the depression of RNA synthesis 3 h after a single dose of galactosamine (Fig. 10). Nucleolar fragmentation has been reversed by uridine as late as 4 h after giving galactosamine (SHINOZUKA et al., 1973b). The reversibility of inhibition of macromolecular synthesis appears essential for reversal or prevention of the processes leading to cell death.

Variation of the Deficit Period by Repeated Small Doses of D-Galactose Analogs

A single galactosamine dose of about 0.6 mmol/kg of body weight causes severe deficiency of UTP and UDP-hexoses and markedly inhibits the induction of tyrosine aminotransferase requiring RNA and protein synthesis (REYNOLDS and REUTTER, 1973). This dose, however, is insufficient to cause liver-cell necrosis because the duration of UTP deficiency under these conditions is too short. However, when several doses of 0.43 mmol/kg were injected at 4-h intervals, signs of hepatitis and cell necrosis became evident (KEPPLER et al., 1968).

An analogous observation was made when 2-deoxy-D-galactose was given together with 6-azauridine (DECKER et al., 1971). A single injection of this combination depressed the hepatic UTP content to less than 30% of normal for about 3 h and a minimum content of 0.02 µmol/g was measured. This deficit period is insufficient to induce cell necrosis and a significant elevation of plasma enzyme activities (Table 3). However, when the period of UTP deficiency was extended by repeating the dose at 2-h or 4-h intervals, cell necrosis occurred (DECKER et al., 1971) (Table 3).

Alteration of the Metabolite Deficit Period by Changes in Hepatic Enzyme Activities

Changes in enzyme activities of UDP-sugar formation or of uridine phosphate synthesis will markedly affect the degree of trapping and the deficiency of uridine phosphates. This situation has been studied particularly with regard to the enzymes of *de novo* pyrimidine biosynthesis (DECKER et al., 1973). Inhibition of orotidine 5'-monophosphate (OMP) decarboxylase by 6-azauridine, which is phosphorylated to 6-azaUMP in liver, greatly inhibits the compensatory increase of uridylate biosynthesis and thereby intensifies the deficiency of UTP and UDP-glucose induced by galactosamine (KEPPLER et al., 1970b) or by 2-deoxy-D-galactose (DECKER et al., 1971).

Several physiologic or pathologic conditions are associated with increased activities of the enzymes of *de novo* pyrimidine biosynthesis, and therefore with better compensation for uridylate trapping (DECKER and KEPPLER, 1972; PAUSCH et al., 1972a, c; DECKER et al., 1973). This relationship has been demonstrated in rats pretreated with galactosamine or orotate (DECKER et al., 1971; PAUSCH et al., 1972b, c), in regenerating liver (BRESNICK, 1965; PAUSCH et al., 1972a, c), and in 10-day-old rats (PAUSCH et al., 1972a; DECKER and KEPPLER, 1972) (Figs. 14, 15).

Daily injections of galactosamine continued for several weeks produce not only signs of progressive chronic hepatitis but also changes in the response of the liver to the amino sugar. With the liver in this adapted state, an injection of 1.16 mmole of galactosamine results in only minimal inflammatory infiltration and insignificant rises in plasma enzyme levels (LESCH et al., 1970a). Repeated injections of galactosamine caused the activities of the enzymes of uridylate biosynthesis to increase, and the basal uridylate levels were elevated (LESCH et al., 1970a) (Fig. 16). It is not clear whether this adaptation is the result of an inductive process in persisting hepatocytes or merely the consequence of continuing liver-

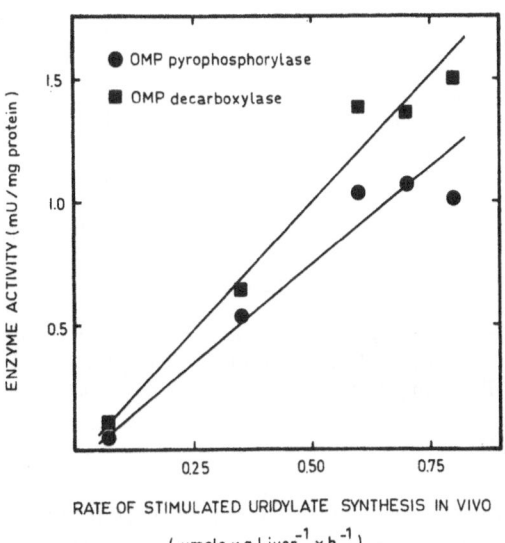

Fig. 14. Correlation of the activities of orotidine 5'-monophosphate (OMP) pyrophosphorylase and OMP decarboxylase with the rate of stimulated uridylate synthesis *in vivo*. The points represent (from left to right) livers of galactosamine-treated guinea pigs, adult, partially hepatectomized, galactosamine-adapted and neonatal rats. (From DECKER et al., 1973)

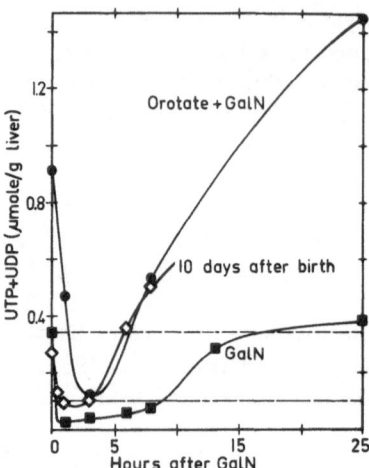

Fig. 15. Diminished uridine phosphate deficit period in rat liver pretreated with orotate and in young rats 10 days after birth. The normal content of UTP + UDP and the 30% level as compared to controls is indicated. (From Decker and Keppler, 1972)

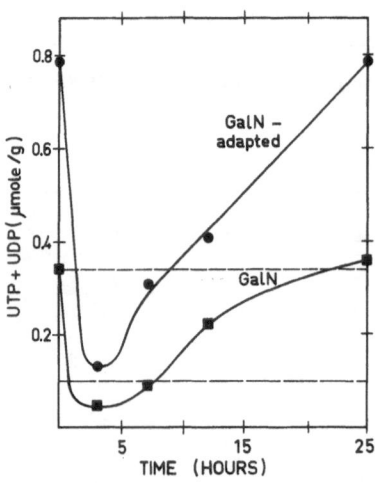

Fig. 16. Diminished uridine phosphate deficit period and increased uridine phosphate contents in the liver of rats pretreated with galactosamine for 3 months (Lesch et al., 1970a). The normal content of UTP + UDP and the level at 30% of controls are indicated by dashed lines. (From Decker et al., 1971)

cell regeneration. In any case, the existence of a galactosamine-refractory state of the liver underscores the importance of enzyme activity correlations in regulating fluxes and levels of metabolites under conditions of "metabolic stress".

Enhanced levels of enzymes of the *de novo* pyrimidine nucleotide biosynthetic pathway are consistently found in fast-growing tissues. This enhanced capacity for uridylate synthesis is, in part, responsible for the galactosamine resistance found in neonatal and regenerating livers (REUTTER et al., 1970); in both cases the liver is in a state of high mitotic activity. Up to about 20 days of age rats do not develop liver injury after the equivalent single dose (per body weight) of galactosamine which proves harmful to older animals; the same phenomenon was observed in regenerating liver between 2 and 6 days after partial hepatectomy. However, if the uridylate deficiency period is extended by repeated galactosamine injections, hepatitis-like damage develops even in the regenerating liver (REUTTER, 1974).

This observation is of great importance as it emphasizes the significance of the uridylate deficiency period and its relationship to the turnover of cell constituents. As pointed out above, uridylate trapping by galactosamine results in severely reduced synthesis of macromolecules such as RNA, proteins, and perhaps heteroglycans. It is safe to assume that under these conditions the turnover of such compounds is shifted toward a relative predominance of degradation with a consequent net loss of vital cell constituents. This process appears to be reversible provided normal synthetic rates are resumed before a critical imbalance is reached. The normal liver cell of an adult rat arrives at this "point of no return" after about 3 h of galactosamine action. It is obvious that this critical time interval depends largely on the rates of the degradative processes, which are markedly slowed down in fast-growing tissues; this has been clearly demonstrated for protein turnover in regenerating liver (SCORNIK, 1972). Therefore, regenerating liver will take longer to reach the critical minimum level of vital cell components during the period of depressed biosynthetic processes. This means, of course, that the metabolite deficit period will be extended under conditions of attenuated turnover. Such interdependence satisfactorily explains the requirement for a prolonged uridylate deficiency period in galactosamine-adapted, neonatal, and regenerating liver.

This view is supported by studies on the hepatic effects of α-amanitin, one of the toxic principles of *Amanita phalloides* (WIELAND, T. and WIELAND, O., 1959, 1972). This octapeptide has been shown to depress RNA synthesis by inhibiting extranucleolar RNA polymerase (STIRPE and FIUME, 1967). A single administration of this compound exerts its effect in rat liver for a limited period of time, which is insufficient to cause irreversible liver damage as judged by ultrastructural examinations after 24 h (MARINOZZI and FIUME, 1971). However, if α-amanitin is administered repeatedly, cell necrosis develops in rat liver (MARINOZZI and FIUME, 1971). Obviously, in this instance, too, the development of irreversible cell injury requires a minimum period of failing RNA synthesis. This is due to inhibition of RNA polymerase in the case of α-amanitin and to deficiency of its substrate, UTP, after galactosamine administration.

Another instance of the dependence of a pathogenetic process on enzyme activity correlations — and hence on primarily kinetic parameters — is the difference in the susceptibility of various species to galactosamine. Among the aminals subjected to galactosamine treatment, mice were least affected by the amino sugar (KEPPLER et al., 1968); it is difficult to observe pathologic changes in these animals even at higher than usual galactosamine doses. On the other hand, guinea pigs are very severely affected by the standard galactosamine treatment. Furthermore, as already mentioned, they are the only species so far found to suffer liver-cell

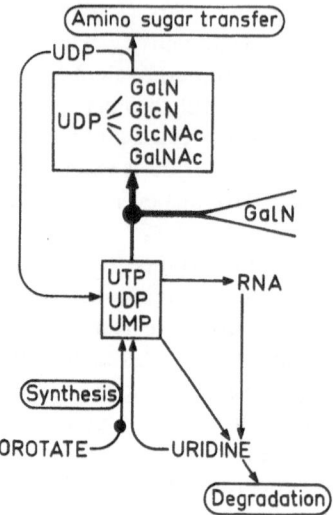

Fig. 17. Factors involved in the induction of uridine triphosphate deficiency by galactosamine and its reversal by pyrimidine nucleotide precursors. For abbreviations, see legends to Figs. 3 and 10. (From KEPPLER et al., 1974)

damage as a result of the administration of 2-deoxy-D-galactose, which is known to trap uridylate about four times more slowly than galactosamine. Measurements of the capacity of total uridylate biosynthesis and of the activities of OMP pyrophosphorylase and OMP decarboxylase revealed that these are highest in mice, intermediate in rats, and lowest in guinea pigs (Fig. 14) (PAUSCH et al., 1972a; DECKER et al., 1973).

Clearly, the extent of uridylate deficiency and the consequent susceptibility to the sugar analogs must depend on the rate of formation and accumulation of the relevant UDP-sugar derivative relative to the rate of uridine phosphate production, both by *de novo* synthesis and by regeneration (Fig. 17).

In the case of galactosamine-induced uridylate trapping, the rate of formation of UDP-galactosamine is determined by the activity of UDP-glucose:galactose-1-phosphate uridylyltransferase as well as by the concentrations of galactosamine-1-phosphate as substrate and UDP-glucose as cosubstrate. In rats, the rate of the galactosamine → UDP-galactosamine conversion *in vivo* exceeds net uridylate synthesis by about 5:1, resulting in a pronounced uridylate deficiency.

Compared with the effects of galactosamine and 2-deoxy-D-galactose, the galactose-induced uridine phosphate deficiency is short-lived and levels of UTP and UDP-glucose do not fall below 30 to 35% of the controls (KEPPLER et al., 1970d). Here, the rate of UDP-glucose formation by UDP-galactose 4'-epimerase and of regeneration of uridylate from UDP-glucose, e.g. by glycogen synthase, is sufficient to counteract the uridylate trapping action and to maintain uridine phosphate levels above the critical concentration.

The steady-state concentrations of the UDP-sugar derivatives depend on the rate of their formation from UTP or UDP-glucose relative to their turnover,

Primary effects: Galactosamine
                                                 ↓
                            Galactosamine-1-phosphate in liver
                                                 ↓
                  Formation and accumulation of UDP-derivatives of galactosamine
                                                 ↓
Primary biochemical lesion: Depletion of hepatic UTP, UDP-glucose, UDP-galactose
                                                 ↓
Secondary lesions: Depression of uracil nucleotide-dependent biosyntheses of macromolecules
                    (ribonucleic acids, proteins, glycoproteins, glycolipids, glycogen)
                                                 ↓
                            Organelle injury in viable cells
                                                 ↓
                               Necrosis of liver cells

Fig. 18. Galactosamine-induced sequence of events. (From DECKER and KEPPLER, 1972)

including sugar transferase reactions. The regeneration of the UDP moiety in sugar transferase reactions may be impaired for two reasons: (a) The UDP-sugar analog accumulated is a poor substrate, if at all, for the respective transferase; (b) the deficiency of UDP-glucose, and in many instances of UDP-galactose as well, negatively affects the transfer of N-acetyl-hexosamines and hexosamines to glycoproteins and glycolipids.

Enzyme activity correlations between biosynthetic and degradative processes determine the turnover of cell constituents. The impact of these correlations on cell growth and tumor promotion has been stressed repeatedly (WEBER et al., 1971). The examples presented above also demonstrate that the relationships between the activities of interconnected synthetic pathways is an important para-meter for the homeostasis of cell metabolism. The quantitative assessment of such relationships, both in the normal state and under pathologic conditions, can be instrumental in elucidating the biochemical mechanism of a pathogenetic sequence (Fig. 18).

The Concept of the "Metabolite Deficit Period" as an Approach to Biochemical Pathology

The evidence obtained from studies of galactosamine action on hepatocytes emphasizes the outstanding importance of the time factor in pathogenesis. Changes in the molecular composition of cells cannot be properly evaluated unless they are defined with regard not only to their subcellular distribution, but also to their quantity and duration. Consideration of these parameters is necessary for an understanding of the operation of cellular metabolism and for an evaluation of the consequences or any deviation.

It is empirically known, and also evident from the many experiments men-tioned above, that each of these three parameters can undergo limited alteration and fluctuation without inflicting irreparable damage on the cell. This biological tolerance introduces some ambiguity into the definition of the normal steady state of a cell in molecular terms. Many factors, of both genetic and environmental origin, constantly produce such fluctuations; endogenous rhythms as well as diet-, age- and activity-dependent changes in metabolic patterns supply examples of physiologically tolerable deviations from a "normal state" which can only be defined as a statistical average in a given set of conditions.

Even if rather large deviations do occur, they may not be fatal to the afflicted cell unless they persist for a certain period of time. Blockade of cellular respiration by anoxia, or depletion of the all-important ATP level by ethionine administration will drastically curtail many vital metabolic functions. These disturbances are reparable and the cell is able to resume its normal activity after a time, provided the deficiency period is not extended beyond a certain limit. The significance of the deficiency period was very clearly demonstrated in the uridine reversal experiments (p. 96). In this case, it was possible to measure the time interval within which the hepatocytes could tolerate a critical deficit of certain metabolites without suffering irreversible damage, although severe lesions of structural components (e.g. nucleoli) and disturbance of metabolic functions (RNA and protein synthesis) were already present.

It has also emerged from these studies that the quantitative aspects of the deficiency itself and of the metabolic and functional consequences are intimately connected with enzyme activity patterns. The susceptibility of an organ or organism to a given toxic agent is determined by activity correlations of certain enzymes, e.g. of uridylate biosynthesis and regeneration versus galactosamine flux through the galactose pathway. Furthermore, the condition of the cell itself, its age and previous experience will influence the turnover and the activity patterns of enzymes. This, in turn, will affect the deficit period and its effects on secondary processes. Examples of this interdependence include the galactosamine resistance of galactosamine-adapted, neonatal, and regenerating liver.

It appears that fatal cell injury requires

(a) that the deviation from the normal state exceeds a threshold limit (band width of allowed fluctuations) *and*

(b) that this deficiency persists over a defined minimum period of time. This concentration-time profile of a metabolic aberration is called the *Metabolite Deficit Period*. It is proposed that the quantitative determination of this profile will lead on to the elucidation of the critical factors in pathogenesis; it may also be helpful in devising strategies to prevent the biochemical lesion from proceeding into an irreversible state. It is, of course, necessary to define the Metabolite Deficit Period for each injurious agent within the metabolic frame of the target cell. Once this period has been established quantitatively, it should be possible to predict the reversibility or lethality of the cell damage elicited by toxic agents and the effects of curative or preventive measures.

References

Anukarahanonta, T., Shinozuka, H., Farber, E.: Inhibition of protein synthesis in rat liver by D-galactosamine. Res. Commun. Chem. Pathol. and Pharmacol. **5**, 481–491 (1973).

Ballard, F. J.: Purification and properties of galactokinase from pig liver. Biochem. J. **98**, 347–352 (1966).

BATES, C. J., ADAMS, W. R., HANDSCHUMACHER, R. E.: Control of the formation of uridine diphospho-N-acetylhexosamine and glycoprotein synthesis in rat liver. J. Biol. Chem. 241, 1705–1712 (1966).

BAUER, CH., BACHMANN, W., REUTTER, W.: Studies on galactosamine hepatitis: Determination of galactosamine metabolites in the developing rat liver. Hoppe-Seyler's Z. Physiol. Chem. 353, 1053–1058 (1972).

BAUER, CH., REUTTER, W.: Inhibition of uridine diphosphoglucose dehydrogenase by galactosamine-1-phosphate and UDP-galactosamine. Biochim. Biophys. Acta 293, 11–14 (1973).

BEKESI, J. G., WINZLER, R. J.: The effect of D-glucosamine on the adenine and uridine nucleotides of sarcoma 180 ascites tumor cells. J. Biol. Chem. 244, 5663–5668 (1969).

BERTOLI, D., SEGAL, S.: Developmental aspects and some characteristics of mammalian galactose-1-phosphate uridyltransferase. J. Biol. Chem. 241, 4023–4029 (1966).

BRESNICK, E.: Early changes in pyrimidine biosynthesis after partial hepatectomy. J. Biol. Chem. 240, 2550–2556 (1965).

BURCH, H. B., MAX, P., JR., CHYU, K., LOWRY, O. H.: Metabolic intermediates in liver of rats given large amounts of fructose or dihydroxyacetone. Biochem. Biophys. Res. Commun. 34, 619–626 (1969).

CHAMBON, P.: Personal communication (1972).

COLOMBI, A., THÖLEN, H., ENGELHARDT, G., HECHT, Y., KOLLER, F.: Gerinnungsfaktoren als Parameter des Leberzellschadens bei akuter Hepatitis. In: K. BECK (Ed.), Ikterus, pp. 323–326. Stuttgart: F. K. Schattauer 1968.

CREASEY, W. A., HANDSCHUMACHER, R. E.: Purification and properties of orotidylate decarboxylases from yeast and rat liver. J. Biol. Chem. 236, 2058–2063 (1961).

CUATRECASAS, P., SEGAL, S.: Mammalian galactokinase, developmental and adaptive characteristics in the rat liver. J. Biol. Chem. 240, 2382–2388 (1965).

DECKER, K., KEPPLER, D.: Galactosamine-induced liver injury. In: H. POPPER and F. SCHAFFNER (Eds.), Progress in Liver Diseases, Vol. 4, pp. 183–199. New York: Grune & Stratton 1972.

DECKER, K., KEPPLER, D., PAUSCH, J.: The regulation of pyrimidine nucleotide level and its role in experimental hepatitis. Advances in Enzyme Regulation 11, 205–230 (1973).

DECKER, K., KEPPLER, D., RUDIGIER, J., DOMSCHKE, W.: Cell damage by trapping of biosynthetic intermediates. The role of uracil nucleotides in experimental hepatitis. Hoppe-Seyler's Z. Physiol. Chem. 352, 412–418 (1971).

DE RITIS, F., COLTORTI, M., GIUSTI, G.: Serum and liver transaminase activities in experimental virus hepatitis in mice. Science 124, 32 (1956).

DE WULF, H., STALMANS, W., HERS, H.-G.: The effect of glucose and of a treatment by glucocorticoids on the activation in vitro of liver glycogen synthetase. Eur. J. Biochem. 15, 1–8 (1970).

FARBER, J. L., GILL, G., KONISHI, Y.: Prevention of galactosamine-induced liver cell necrosis by uridine. Am. J. Pathol. 72, 53–62 (1973).

FISCHER, W., WEIDEMANN, G.: Nachweis des Leloir-Weges in Pflanzen mit 2-Desoxy-D-galaktose. Biochim. Biophys. Acta 93, 677–680 (1964a).

FISCHER, W., WEIDEMANN, G.: Die Umsetzung von 2-Desoxy-D-galaktose im Stoffwechsel. II. Identifizierung der phosphorylierten Zwischenprodukte. Hoppe-Seyler's Z. Physiol. Chem. 336, 206–218 (1964b).

GERLACH, U., HIBY, W.: Sorbit-Dehydrogenase. In: Methoden der enzymatischen Analyse (H. U. BERGMEYER, Ed.), 2nd ed., pp. 527–533. Weinheim: Verlag Chemie 1970.

GRÜN, M., BRUNSWIG, D., MERSCH-BAUMERT, K., CONRADT, M., HÖRDER, M. H., LAUN, A., LIEHR, H., KRAUSS, H., RICHTER, E.: Gerinnungsstörungen bei der „Galaktosamin-Hepatitis" der Ratte. Acta Hepato-Gastroenterol. 19, 103–108 (1972).

HÄNNINEN, D., MARNIEMI, J.: Inhibition of glucuronide synthesis by physiological metabolites in liver slices. FEBS Letters 6, 177–181 (1970).

KEPPLER, D.: Hépatite expérimentale à la D-galactosamine. Ann. Gastroent. Hepatol. 9, 211–221 (1973).

KEPPLER, D., BISCHOFF, E., DECKER, K.: The role of uridine nucleotides in galactosamine hepatitis. 7th Congr. Clin. Chem., Geneva/Evian 1969; Vol. 3, pp. 456–460. Basel: Karger 1970c.

KEPPLER, D., DECKER, K.: Studies on the mechanism of galactosamine hepatitis: Accumulation of galactosamine-1-phosphate and its inhibition of UDP-glucose pyrophosphorylase. Eur. J. Biochem. **10**, 219–225 (1969).

KEPPLER, D., DECKER, K.: Der Wirkungsmechanismus von D-Galaktosamin in der Leber. Verh. Deutsch. Ges. Inn. Med. **77**, 1179–1182 (1971).

KEPPLER, D., DECKER, K.: UTP depletion induced by sugars: Reversibility and early consequences in liver. Abstr. Commun. Meet. Fed. Europ. Biochem. Soc. **8**, 1043 (1972).

KEPPLER, D., FRÖHLICH, J., REUTTER, W., WIELAND, O., DECKER, K.: Changes in uridine nucleotides during liver perfusion with D-galactosamine. FEBS Letters **4**, 278–280 (1969).

KEPPLER, D., HÜBNER, G.: Liver injury induced by 2-deoxy-D-galactose. Exptl. Mol. Pathol. **19**, 365–377 (1973).

KEPPLER, D., LESCH, R., REUTTER, W., DECKER, K.: Experimental hepatitis induced by D-galactosamine. Exptl. Mol. Pathol. **9**, 279–290 (1968).

KEPPLER, D., PAUSCH, J., DECKER, K.: Selective uridine triphosphate deficiency induced by D-galactosamine in liver and reversed by pyrimidine nucleotide precursors. Effect on ribonucleic acid synthesis. J. Biol. Chem. **249**, 211–216 (1974).

KEPPLER, D., RUDIGIER, J., BISCHOFF, E., DECKER, K.: The trapping of uridine phosphates by D-galactosamine, D-glucosamine, and 2-deoxy-D-galactose. A study on the mechanism of galactosamine hepatitis. Eur. J. Biochem. **17**, 246–253 (1970b).

KEPPLER, D., RUDIGIER, J., DECKER, K.: Trapping of uridine phosphates by D-galactose in ethanol-treated liver. FEBS Letters **11**, 193–196 (1970d).

KEPPLER, D., RUDIGIER, J., DECKER, K.: Enzymic determination of uracil nucleotides in tissues. Anal. Biochem. **38**, 105–114 (1970e).

KEPPLER, D., RUDIGIER, J., REUTTER, W., LESCH, R., DECKER, K.: Orotate prevents galactosamine hepatitis. Hoppe-Seyler's Z. Physiol. Chem. **351**, 102–104 (1970a).

KEPPLER, D., SMITH, D. F.: Nucleotide contents of ascites hepatoma cells and their changes induced by D-galactosamine. Cancer Res. **34**, 705–711 (1974).

KOFF, R. S., GORDON, G., SABESIN, S. M.: D-Galactosamine hepatitis. I. Hepatocellular injury and fatty liver following a single dose. Proc. Soc. Exptl. Biol. Med. **137**, 696–701 (1971).

KORNFELD, S., KORNFELD, R., NEUFELD, E. F., O'BRIEN, P. J.: The feedback control of sugar nucleotide biosynthesis in liver. Proc. Natl. Acad. Sci. U. S. **52**, 371–379 (1964).

KUHN, R., LEPPELMANN, J. H., FISCHER, H.: Stoffwechselversuche mit radioaktiv markiertem D-Glucosamin und D-Galaktosamin. Liebigs Ann. Chem. **620**, 15–20 (1959).

LESCH, R., BAUER, CH., REUTTER, W.: The development of cholangiofibrosis and hepatomas in galactosamine-induced cirrhotic rat livers. Virchow's Arch. Abt. B Zellpath. **12**, 285–289 (1973).

LESCH, R., KEPPLER, D., REUTTER, W., RUDIGIER, J., OEHLERT, W., DECKER, K.: Entwicklung einer experimentellen Leberzirrhose durch D-Galactosamin. Histologische, biochemische und autoradiographische Untersuchungen an Ratten. Virchow's Arch. Abt. B Zellpath. **6**, 57–71 (1970a).

LESCH, R., REUTTER, W., KEPPLER, D., DECKER, K.: Liver restitution after acute galactosamine hepatitis: Autoradiographic and biochemical studies in rats. Exptl. Mol. Pathol. **12**, 58–69 (1970b).

LEVINE, R. L., HOOGENRAAD, N. J., KRETCHMER, N.: Regulation of activity of carbamoyl phosphate synthetase from mouse spleen. Biochemistry **10**, 3694–3699 (1971).

MALEY, F.: The synthesis of UDP-galactosamine and UDP-N-acetylgalactosamine. Biochem. Biophys. Res. Commun. **39**, 371–378 (1970).

MALEY, F., MALEY, G. F.: The enzymic conversion of glucosamine to galactosamine. Biochim. Biophys. Acta **31**, 577–578 (1959).

MALEY, F., TARENTINO, A. L., McGARRAHAN, J. F., DELGIACCO, R.: The metabolism of D-galactosamine and N-acetyl-D-galactosamine in rat liver. Biochem. J. **107**, 637–644 (1968).

MARINOZZI, V., FIUME, L.: Effects of α-amanitin on mouse and rat liver cell nuclei. Exper. Cell Res. **67**, 311–322 (1971).

MARNIEMI, J., NURMINEN, J., HÄNNINEN, O.: Studies on the action of D-galactose and D-galactosamine on the glucuronide synthesis in the rat tissues. Int. J. Biochem. **4**, 227–233 (1973).

MEDLINE, A., SCHAFFNER, F., POPPER, H.: Ultrastructural features in galactosamine-induced hepatitis. Exptl. Mol. Pathol. **12**, 201–211 (1970).

MERSMANN, H. J., SEGAL, H. L.: An on–off mechanism for liver glycogen synthetase activity. Proc. Natl. Acad. Sci. U. S. **58**, 1688–1695 (1967).

MÖHR, J. R., MATTENHEIMER, H., HOLMES, A. W., DEINHARDT, F., SCHMIDT, F. W.: Enzymology of experimental liver disease in Marmoset monkeys. Enzyme **12**, 161–179 (1971).

MONIER, D., WAGLE, S. R.: Studies on gluconeogenesis in galactosamine-induced hepatitis. Proc. Soc. Exptl. Biol. Med. **136**, 377–380 (1971).

MÜLLER-BERGHAUS, G., REUTER, C.: Disseminated intravascular coagulation in galactosamine-induced experimental hepatitis. Thrombosis Res. **1**, 473–486 (1972).

PAUSCH, J., KEPPLER, D., DECKER, K.: Activity and distribution of the enzymes of uridylate synthesis from orotate in animal tissues. Biochim. Biophys. Acta **258**, 395–403 (1972a).

PAUSCH, J., KEPPLER, D., DECKER, K.: Increased activities of hepatic orotidine 5′-phosphate pyrophosphorylase and orotidine 5′-phosphate decarboxylase induced by orotate. FEBS Letters **20**, 330–332 (1972b).

PAUSCH, J., KEPPLER, D., DECKER, K.: Zur Bedeutung der Aktivität von Enzymen der Pyrimidinbiosynthese für die Entwicklung der Galaktosaminschädigung der Leber. Verh. Deutsch. Ges. Inn. Med. **78**, 1375–1378 (1972c).

REUTTER, W., BACHMANN, W.: Galaktosaminhepatitis: Schädigung der Plasmamembran der Leberzelle nach Gabe von D-Galaktosamin. Verh. Deutsch. Ges. Inn. Med. **77**, 1177–1178 (1971).

REUTTER, W., BAUER, CH., BACHMANN, W., LESCH, R.: Über die primäre und sekundäre biochemische Antwort der Leber nach Gabe von D-Galaktosamin beim Zustandekommen der Galaktosaminhepatitis. Verh. Deutsch. Ges. Inn. Med. **79**, 927–930 (1973).

REUTTER, W., BAUER, CH., BACHMANN, W., LESCH, R.: The galactosamine-refractory regenerating rat liver. *In*: R. LESCH and W. REUTTER (Eds.), Liver regeneration after experimental injury. New York: Intercontinental Medical Book Corp., in press.

REUTTER, W., BAUER, CH., LESCH, R.: On the mechanism of action of galactosamine: Different response to D-galactosamine of rat liver during development. Die Naturwissenschaften **57**, 674–675 (1970).

REUTTER, W., KEPPLER, D., LESCH, R., DECKER, K.: Zum Glykoproteidstoffwechsel bei der Galaktosamin-induzierten Hepatitis. Verh. Deutsch. Ges. Inn. Med. **75**, 363–365 (1969).

REYNOLDS, R. D., REUTTER, W.: Inhibition of induction of rat liver tyrosine aminotransferase by D-galactosamine. Role of uridine triphosphate. J. Biol. Chem. **248**, 1562–1567 (1973).

SCHARNBECK, H., SCHAFFNER, F., KEPPLER, D., DECKER, K.: Ultrastructural studies on the effect of choline orotate on galactosamine induced hepatic injury in rats. Exptl. Mol. Pathol. **16**, 33–46 (1972).

SCHMIDT, E.: Glutamat-Dehydrogenase, UV-Test. *In*: Methoden der enzymatischen Analyse (H. U. BERGMEYER, Ed.), 2nd Ed., pp. 607–613. Weinheim: Verlag Chemie 1970.

SCHOLTISSEK, C.: Detection of an unstable RNA in chick fibroblasts after reduction of the UTP pool by glucosamine. Eur. J. Biochem. **24**, 358–365 (1971).

SCORNIK, O. A.: Decreased *in vivo* disappearance of labelled liver protein after partial hepatectomy. Biochem. Biophys. Res. Commun. **47**, 1063–1066 (1972).

SHINOZUKA, H., FARBER, J. L., KONISHI, Y., ANUKARAHANONTA, T.: D-Galactosamine and acute liver cell injury. Fed. Proc. **32**, 1516–1526 (1973a).

SHINOZUKA, H., MARTIN, J. T., FARBER, J. L.: The induction of fibrillar nucleoli in rat liver cells by D-galactosamine and their subsequent re-formation into normal nucleoli. J. Ultrastruct. Res. **44**, 279–292 (1973b).

SIEBERT, G.: The biochemical environment of the mammalian nucleus. Sub-Cell. Biochem. **1**, 277–292 (1972).

SMITH, P. C., KNOTT, C. E., TREMBLAY, G. C.: Detection of the feedback control of pyrimidine biosynthesis in slices of several rat tissues. Biochem. Biophys. Res. Commun. **55**, 1141–1146 (1973).

STIRPE, F., FIUME, L.: Studies on the pathogenesis of liver necrosis by α-amanitin. Effect of α-amanitin on ribonucleic acid synthesis and on ribonucleic acid polymerase in mouse liver nuclei. Biochem. J. **105**, 779–782 (1967).

THORN, W.: Organstoffwechsel und Organfunktion. *In*: H. BARTELHEIMER, W. HEYDE, W. THORN (Eds.), „D-Glucose und verwandte Verbindungen in Medizin und Biologie", p. 378–425. Stuttgart: Enke Verlag 1966.

WALKER, D. G., KHAN, H. H.: Some properties of galactokinase in developing rat liver. Biochem. J. **108**, 169–175 (1968).

WEBER, G., QUEENER, S. F., FERDINANDUS, J. A.: Control of gene expression in carbohydrate, pyrimidine and DNA metabolism. Advances in Enzyme Regulation **9**, 63–95 (1971).

WHITE, B. N., SHETLAR, M. R., SHURLEY, H. M., SCHILLING, J. A.: Incorporation of D-[1-^{14}C]-galactosamine into serum proteins and tissues of the rat. Biochim. Biophys. Acta **101**, 259–266 (1965).

WIELAND, T., WIELAND, O.: Chemistry and toxicology of the toxins of *Amanita phalloides*. Pharmacol. Rev. **11**, 87–107 (1959).

WIELAND, T., WIELAND, O.: The toxic peptides of *amanita* species. In: S. KADIS, A. CIEGLER, S. J. AIL (Eds.), Microbial Toxins, Vol. 8, pp. 249–280. New York and London: Academic Press 1972.

Rev. Physiol. Biochem. Pharmacol., Vol. 71
© by Springer Verlag 1974

Recent Concepts of Intestinal Fat Absorption *

Robert K. Ockner and Kurt J. Isselbacher **

Contents

I. Introduction

Although the subject of fat absorption has been considered in several earlier
reviews (Senior, 1964; Isselbacher, 1967; Johnston, 1968), it seems appropriate
to re-examine this complex process in the context of more recent progress in
the area itself and in related aspects of lipid metabolism and physiology. In the
past several years, newer concepts have been advanced in the physical chemistry
of lipids, the physiology of bile salts, and in several aspects of the formation

* Supported in part by Research Grants AM-13328 and AM-14795, and by Research Career
Development Award AM-36586, from the National Institutes of Health, Bethesda, Maryland,
U.S.A.
** Departments of Medicine, University of California, School of Medicine, San Francisco,
Ca./USA and Harvard Medical School and The Massachusetts General Hospital, Boston,
Mass./USA.

and metabolism of lipoproteins that bear directly on the process of fat absorption. Increased understanding of the physiological function of the lipoprotein peptides has led to a fuller appreciation of the relationship between the lipoproteins formed in the intestine (both from endogenous and exogenous lipids) and those that circulate in plasma.

Some previously unresolved questions regarding fat absorption can now be explained, whereas the answers to others are still outstanding. It is noteworthy that, of those findings of the past 10 to 15 years which in their own time were considered to represent significant advances, few have failed to stand the test of time. Virtually all, however, have required modification in the light of recent information.

In the present discussion, emphasis is placed on the absorption of long-chain fatty acids from intestinal lumen to plasma and attention is directed toward the consideration of newer concepts and, where appropriate, a re-interpretation of older concepts. Although the review is basic in its overall orientation, clinical phenomena are alluded to insofar as they contribute to an understanding of normal and abnormal physiology and biochemistry. The absorption of medium-chain fatty acids, sterols, or phospholipids is not considered in significant detail, and no discussion of pancreatic lipolysis is included.

Even in regard to the areas selected for emphasis, we have not attempted to be encyclopedic. On the contrary, in order to present a reasonably concise and cohesive overview of several major areas in which there has been considerable recent progress, reference to some significant publications may have been omitted. We regret this, but hope that the concepts discussed may serve to stimulate additional inquiry into these and other aspects of intestinal function and lipid transport and metabolism.

II. Products of Intraluminal Lipolysis: Physico Chemical Properties and the Bile-Salt Micelle

The reaction of pancreatic lipase with triglyceride results in the formation of fatty acid and 2-monoglyceride. For the most part, these compounds are absorbed as such by the intestinal mucosa, although a small amount of monoglyceride is further hydrolyzed to fatty acid and glycerol (Fig. 9). These products of triglyceride hydrolysis are soluble in water to only a limited extent and, therefore, the physical state in which they exist within the lumen of the gastrointestinal tract is of major importance in the overall process of fat absorption. An understanding of this state, the subject of an excellent recent view by Hofmann and Mekhjian (1973), can best be gained by considering the intrinsic physicochemical properties of the lipids themselves and the nature of their interaction with water and with bile salts.

Lipids of biological importance have been classified by Small (1968, 1970) and by Carey and Small (1970) on the basis of their interaction with water (Fig. 1). Triglycerides, diglycerides and undissociated long-chain fatty acids are

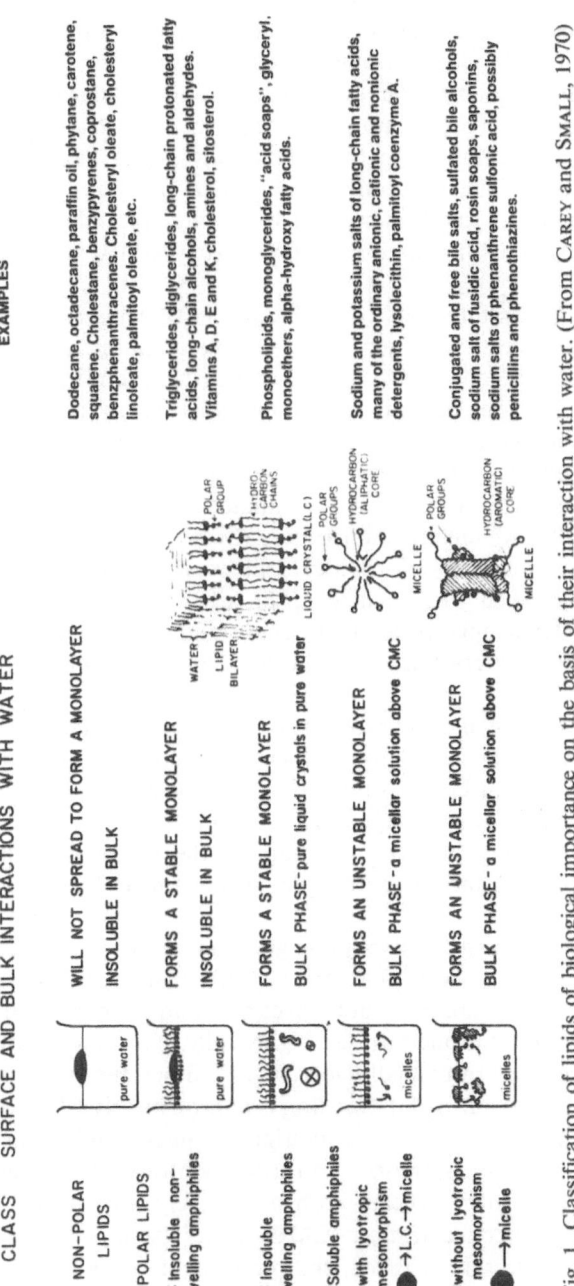

Fig. 1. Classification of lipids of biological importance on the basis of their interaction with water. (From CAREY and SMALL, 1970)

designated "insoluble, non-swelling amphiphiles". These lipids are not only insoluble in water but show virtually no tendency to interact with water in bulk phase. Thus, their admixture to water results in the formation of an unstable emulsion or a two-phase system. In this respect they differ from another class of insoluble lipids (including monoglycerides and phospholipids) which do interact with water in bulk phase by forming (at temperatures above the transition temperature of the lipid) "lamellar liquid crystals". The lamellar liquid crystal results from the penetration of water molecules between polar head groups of adjacent lipid molecules ("swelling"), so that alternating layers (lamellae) of lipid and water are formed. While not constituting a true solution, such liquid crystals do reflect the more intimate interaction with water that characterizes this class of lipids, compared with that exhibited by triglycerides and the other "non-swelling" lipids.

Sodium and potassium salts of long-chain fatty acids are designated, along with bile salts and other detergents, as "soluble amphiphiles". At low concentrations, these compounds exist as monomers in true solution, but as their concentration increases, they begin to form multimolecular aggregates or "micelles". The lowest monomer concentration at which such aggregates are present in solution is termed the "critical micellar concentration". The solubility of long-chain fatty-acid anions, however, is quite limited and at concentrations exceeding approximately 10^{-6} M dimerization occurs (Mukerjee, 1965). Long-chain fatty acids by themselves also may form higher states of aggregation (Smith and Tanford, 1973), but when their concentration exceeds their maximum solubility (a function of pH, temperature, chain length and degree of unsaturation) they exist in bulk phase as crystals or liquid crystals, depending on their transition temperature.

Because triglyceride and the products of its hydrolysis (fatty acids and monoglyceride) are so poorly suited to a stable interaction with an aqueous medium, it is understandable that fat absorption depends, for maximal efficiency, on some mechanism by which these compounds can achieve this stability in bulk water phase, thereby allowing for their exposure to the largest possible area of the intestinal absorptive surface. Bile salts play a fundamental role in this stabilizing process.

The physical, chemical, and physiological properties of bile salts have been reviewed elsewhere (Hofmann and Small, 1967; Carey and Small, 1970). Bile salts are "soluble amphiphiles" but, in contrast to long-chain fatty acids, the physiologically important bile salts are highly soluble in water at pH levels above their pKa, i.e. as nonprotonated acids (Fig. 2). Since the pH normally present in the intestinal lumen during fat digestion is approximately 6.0 to 6.5, the physiologically predominant glycine- and taurine-conjugated bile salts exist almost entirely in this form (i.e. as anions). In further contrast to fatty acids, bile salts at physiological pH form micelles with a relatively high aggregation number. The critical micellar concentration (CMC) varies from 2 to 5 mM depending upon the specific bile salt involved, the presence of other lipids in the system, and the concentration and nature of inorganic salts and cations ("counterions") present. The bile-salt molecules in the micelle exist in dynamic equilibrium with their corresponding monomers in solution.

In addition to the characteristic tendency of bile-salt molecules to form micelles, these compounds also interact with other lipids. "Swelling amphiphiles" such as phospholipid and monoglyceride are readily incorporated into the bile-

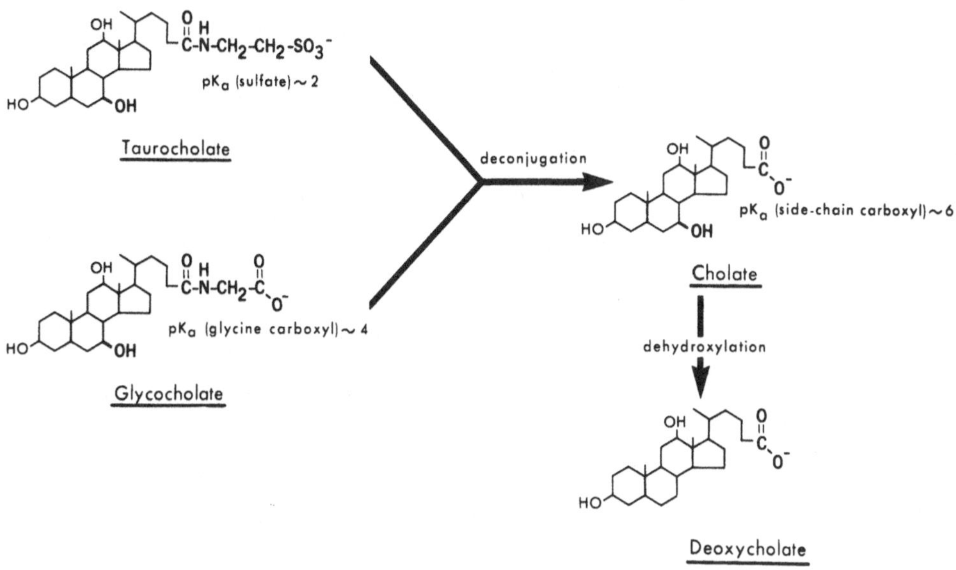

Fig. 2. Metabolism of bile salts by intestinal bacteria

salt micelle, and the resulting "mixed micelle" (i.e. a micelle composed of more than one lipid) is quite stable in water. The capacity of bile salts to incorporate swelling amphiphiles into the micelle is substantial and is of critical significance biologically, not only in the solubilization of monoglyceride in the intestinal lumen, but also in the solubilization of lecithin in bile. The incorporation of swelling amphiphiles (monoglyceride and lecithin) into bile-salt micelles has the additional effect of significantly increasing the size and aggregation number of the micelle itself. This then permits other relatively nonpolar lipids (e.g. cholesterol), which are less readily incorporated into pure bile-salt micelles, to be carried in the hydrophobic interior of the expanded mixed micelle. (Because there is virtually no interaction of bile salts with triglyceride or other lipids of its class; significant quantities of triglyceride are not incorporated into the mixed micelle.)

Long-chain fatty acids also are incorporated readily into bile-salt micelles, especially in the presence of monoglyceride. The resulting mixed bile salt-monoglyceride-fatty acid micelle not only accounts largely for the intraluminal solubilization of the otherwise poorly soluble monoglyceride and fatty acid, but in addition provides the vehicle by which other biologically important lipid-soluble compounds including sterols (e.g. cholesterol and vitamin D) can be stabilized in the aqueous phase of the intestinal contents and thus be absorbed more efficiently.

The foregoing considerations of mixed micelle formation in the intestinal lumen, based largely upon observations using *in vitro* systems and the physico-chemical properties of the individual components of the micelle, are well established. Certain physical properties of the micelle *itself*, however, such as size and structure, are less well understood. Because the essential component in the micelle (i.e. the bile-salt molecule) is not a soap or synthetic detergent with a

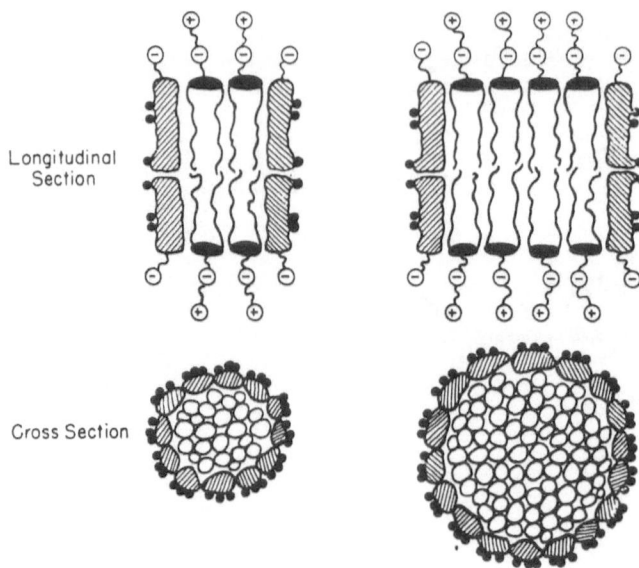

Fig. 3. Schematic representation of the mixed bile salt-lecithin micelle. (From SMALL et al., 1969)

relatively linear molecular configuration (such compounds tend to form spherical micelles), but rather is a structurally complex molecule in which hydrophobic (the steroid nucleus) and hydrophilic (the hydroxyl groups and the side-chain carboxyl or sulfate groups) regions have a relatively planar orientation, it has been difficult to conceptualize the spatial arrangement of individual molecules within the micelle. It has been suggested that, rather than forming perfectly spherical micelles, these compounds form aggregates which more nearly resemble cylinders (SMALL et al., 1969). In this model, bile salts are oriented longitudinally along the surface, and the ends of the cylinder are composed of the hydrophilic groups of the polar lipids (phospholipid, monoglyceride or fatty acid) that may be incorporated into the micelle. A schematic representation of this arrangement has been proposed by SMALL et al. (1969) and is shown for the special case of the bile salt-lecithin micelle (Fig. 3). This model is consistent with the physical properties of the compounds themselves, as well as with data provided by nuclear magnetic resonance spectroscopy.

Efforts have also been made to determine the approximate size of the mixed bile-salt micelle. Unfortunately, as CAREY and SMALL (1970) have pointed out, many of the techniques employed for such estimations were designed primarily for determining the size of *macromolecules* (such as proteins) rather than *multimolecular aggregates* (such as micelles), the components of which exist in equilibrium with corresponding monomers in solution. Nonetheless, results obtained by such techniques, including gel filtration (BORGSTRÖM, 1965; FELDMAN and BORGSTRÖM, 1966), sedimentation equilibrium (LAURENT and PERSSON, 1965), and X-ray diffraction have been remarkably consistent, and suggest that the radius of the mixed bile-salt micelle (considering an equivalent sphere) is of the

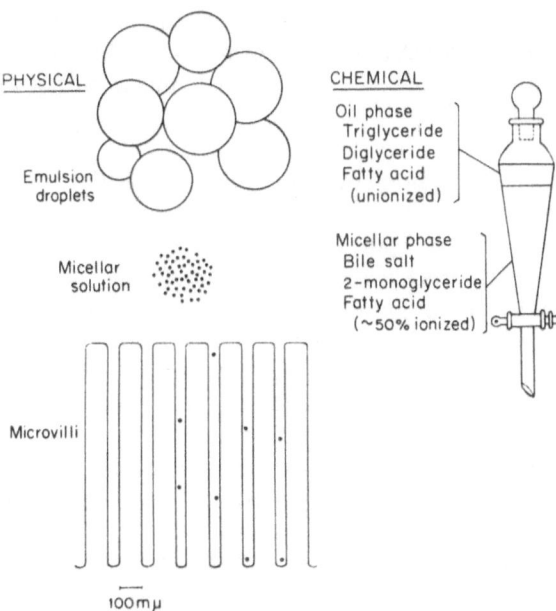

Fig. 4. Separation of phases of small intestinal contents during digestion of a fatty meal (From HOFMANN, 1970)

order of 30 to 60 Å. Because the pH of the intestinal lumen is above the pK_a of the conjugated bile salts and approximates that of the long-chain fatty acids, the micelle has a net negative charge.

A fundamental property of the micelle, which not only is implicit in its defini-tion but may also play a critical role in fat absorption, is the fact that a dynamic equilibrium exists between the individual component of the micelle and its cor-responding monomer (e.g. the bile salt or fatty acid anion) in the bulk water phase. Thus, the micelle cannot be regarded as a "package", the composition of which remains fixed irrespective of changes in the surrounding bilk phase. Rather, as the aqueous (monomer) concentration of one of the micelle com-ponents changes, so will its tendency to enter or leave the micelle. This equilibrium must be recognized and considered in any proposed mechanism by which such components subsequently may enter the intestinal absorptive cell.

A physiological role for bile salt-monoglyceride-fatty acid micelles in the absorption of lipid was established by the landmark experimental studies of HOFMANN and BORGSTRÖM (1962). HOFMANN (1961) presented *in vitro* evidence that fatty acid and monoglyceride could be solubilized by bile salts under condi-tions of ionic strength, temperature, and pH that occur in the intestinal lumen. HOFMANN and BORGSTRÖM later showed (1964), that during the process of fat digestion and absorption in man, an aqueous "micellar phase" could be separated from jejunal contents by high-speed centrifugation. This micellar phase contained bile salt, monoglyceride and fatty acid, but virtually no triglyceride or diglyceride (Fig. 4). The latter lipids ("insoluble, non-swelling amphiphiles") were present in a floating oil layer which probably existed in the intestine as an unstable emulsion. These findings have been confirmed by other investigators, and the existence of a micellar phase in the intestinal lumen during fat absorption is now a well-established fact.

Quantitative aspects of this micellar phase (e.g. the actual concentration of fatty acid, bile salt and monoglyceride contained within it) are less certain. Porter and Saunders (1971) showed that during the digestion and absorption of a fatty meal, significant lipolysis may occur *while luminal samples are being collected and processed*, even if they are heat-treated in order to inactivate pancreatic lipase. These workers also showed that, after centrifugation of intestinal contents, the resulting micellar (aqueous) phase is not homogeneous but establishes a gradient in the centrifuge tube. Consequently, concentrations of lipids in the micellar phase will vary, depending on the region of the tube from which the sample is obtained. This observation does not detract from the significance of the earlier observations of Hofmann and Borgström. However, it does serve to emphasize the need for caution in the design and interpretation of any experiment in which conclusions are to be drawn from measured concentrations of micellar phase lipids in experimental or pathologic circumstances, and bears out the suggestion of Dawson (1967) that such technical problems might require more attention than previously given to them.

Elucidation of the pathophysiology of a number of clinical disorders of fat absorption has shed additional light on the significance of micelle formation during fat absorption. Thus, malabsorption of fat results when intraluminal bile-salt concentrations are reduced to below the CMC (Van Deest et al., 1968; Hofmann and Kern, 1971). This may occur in severe liver disease or bile-duct obstruction, both of which result in an inadequate rate of entry of bile salts into the duodenum. Diminished luminal bile-salt concentrations may also be caused by three general categories of disordered intestinal function.

One condition is generally referred to as the "*intestinal stasis syndrome*". In this disorder, a variety of derangements, including impaired intestinal motility, chronic partial obstruction, or fistula formation, may result in colonization of proximal small intestine by increased numbers of bacteria, some of which (particularly the anaerobic Bacteroides) are capable of deconjugating bile salts (Fig. 2). The resulting loss of the taurine or glycine moiety raises the pK_a of the bile salt molecule to that of the side-chain carboxyl group (i.e. to approximately 6.0) and thereby greatly diminishes the concentration of bile salt which exists in the ionized form at the pH of the intestinal contents. The corresponding increase in unionized (protonated) bile salt leads to increases both in the amount which is passively absorbed nonionic diffusion (Dietschy, 1968) and in the amount which precipitates from solution. The net effect of deconjugation, therefore, is to reduce the intraluminal bile-salt concentration and, secondarily, to diminish micelle formation. In this disorder, it is also possible that deoxycholate, formed by bacterial *dehydroxylation* of cholate (Fig. 2), may contribute to steatorrhea by causing mucosal injury (Dawson and Isselbacher, 1960; Ament et al., 1972).

A second intestinal cause of reduced luminal bile-salt concentration is *decreased ileal bile-salt reabsorption*, due either to ileal disease or the surgical removal or bypass of this portion of the intestine (Hardison and Rosenberg, 1967).

Under these conditions there is an increase in the loss of bile salts from the enterohepatic circulation into the colon and feces; as a result the amount of bile salt returning to the liver via the portal vein is diminished. In compensation, hepatic bile-salt synthesis increases and, if the loss into the colon is not excessive, augmented synthesis may actually be sufficient to maintain luminal bile salt concentrations. However, with greater fecal losses, the maximal bile-salt synthetic rate of which the liver is capable (about 5 times normal) may be inadequate, with the result that total bile-salt pool size falls and luminal bile-salt concentrations drop to below the critical micellar concentration (HOFMANN and POLEY, 1969).

A third, but much less common cause of diminished luminal bile-salt concentration is the *Zollinger-Ellison syndrome* (SHIMODA et al., 1968; ISENBERG et al., 1973). This condition is associated with marked gastric hypersecretion of acid, due to a gastrin-secreting, non-beta islet-cell tumor of the pancreas. Under these circumstances, the increased rate of entry of acid into the duodenum may exceed the buffering capacity of the bicarbonate-rich pancreatic and biliary secretions, with a consequent reduction of luminal pH. If the pH approximates the pK_a of the glycine- ($\sim pH$ 4.0) or taurine-conjugates ($\sim pH$ 2.0), then luminal bile-salt concentration will fall (as in the stasis syndrome) because of increased absorption by nonionic diffusion or intraluminal precipitation.

These clinical examples of fat malabsorption caused by diminished luminal bile-salt concentrations serve to emphasize the importance of micelle formation in fat absorption. However, bile salts are by no means indispensable to this process. Thus, in the presence of complete biliary obstruction or bile fistula, patients and experimental animals alike are able to absorb most (50–70%) of the dietary fat ingested (BORGSTRÖM, 1953; GALLAGHER et al., 1965). Studies of PORTER et al. (1971) suggest that in man fat absorption takes place under these conditions by normal pathways (i.e. triglyceride synthesis and chylomicron formation). However, in experimental animals, fatty acid re-esterification may be defective, with the result that increased amounts of lipid are absorbed as unesterified fatty acids into the portal blood (GALLAGHER et al., 1965). In either case, the formation of mixed bile-salt micelles in the intestinal lumen is appropriately regarded as a mechanism by which the process of fat absorption is facilitated and rendered more complete, but one which does not constitute an absolute requirement for it.

In *summary*, the products of triglyceride hydrolysis (i.e. free fatty acids and 2-monoglyceride) are for practical purposes insoluble in aqueous systems. However, the presence of bile salts in the intestinal contents during fat digestion results in the formation of a multimolecular aggregate, i.e. the mixed bile salt-fatty acid-monoglyceride micelle, which is stable in water, carries a net negative charge, and is in dynamic equilibrium with its more polar components, each of which may exist in monomeric aqueous solution at relatively low concentration. The mechanism by which these products of triglyceride hydrolysis subsequently gain entry into the absorptive epithelium is not fully understood, but must take into account the foregoing physicochemical considerations, the nature of the epithelial surface itself, and the interface between the luminal contents and the absorptive cell surface.

III. Epithelial Uptake of Lipids: The Lumen-Absorptive Cell Interface

A. Structural Considerations

Many aspects of the nature and structure of the intestinal absorptive epithelium have long been known, but more recent advances in the understanding of this unique surface warrant a reconsideration of its relationship to the process of fat absorption (TRIER, 1968, 1969). A particularly striking feature of the absorptive epithelium is the manner in which it presents an enormous surface area to the luminal contents. This is achieved not only by the length of the intestine itself, the reduplications of its mucosal layer (the plicae circularis or valvulae coniventes), and the finger-like projections of the epithelium and lamina propria which constitute the villi, but also of specialized cellular organelles, the "microvilli", which are reduplications of the luminal aspect of the plasma membrane of each absorptive cell. These microvilli (Figs. 5, 6), which collectively comprise the "brush border" of the cell, are approximately 1 μm in length, relative to an overall cell height of 20–25 μm. Their diameter is about 0.1 μm (i.e. 1000 Å), and they are separated from each other by a similar distance (500–1000 Å). It has been estimated that the microvilli effect a 14 to 39-fold increase in the area of the absorptive surface (TRIER, 1968).

The microvilli themselves consist of three recognizable and distinct morphological elements (Fig. 5). The first of these, the plasma membrane, exhibits a typical trilaminar "unit membrane" structure in osmium-fixed sections examined by electron microscopy. It is about 95–115 Å in thickness, whereas the plasma membrane along the lateral and basal aspects of the cell measures 70–80 Å.

The second recognizable anatomic feature of the microvillus is the contained fibrillar material, which is oriented along the long axis of the structure and comprises the "central core"; cross fibrils also occur. The core is believed to provide mechanical support, but may in addition provide motility to the microvilli since it contains contractile proteins such as actin (TILNEY and MOOSEKER, 1971). These fibrils extend within the cell to below the microvillus surface, where they intermesh to form the "terminal web". The latter structure marks the demarcation between the brush border and the apical cytoplasm of the cell, which contains the apical vesicles of the endoplasmic reticulum, i.e. the site of fatty-acid activation and esterification (see below).

The third feature of the microvillus surface, the "glycocalyx" or "fuzz" or "surface coat", has been described more recently (ITO, 1965; TRIER, 1968) and its function is still relatively poorly understood. It is present in all mammalian species, but is most highly developed in man, bat and cat. The glycocalyx is composed of a rich network of glycoproteins which constitute an integral part of the microvillus membrane and cannot be separated from it so long as the cell itself is viable. It has been suggested that this surface coat exerts a protective function by preventing the interaction of large or potentially destructive intraluminal objects (such as bacteria) with the absorptive cell plasma membrane, but this concept has not been established conclusively. It may also contain certain enzymes, such as the disaccharidases.

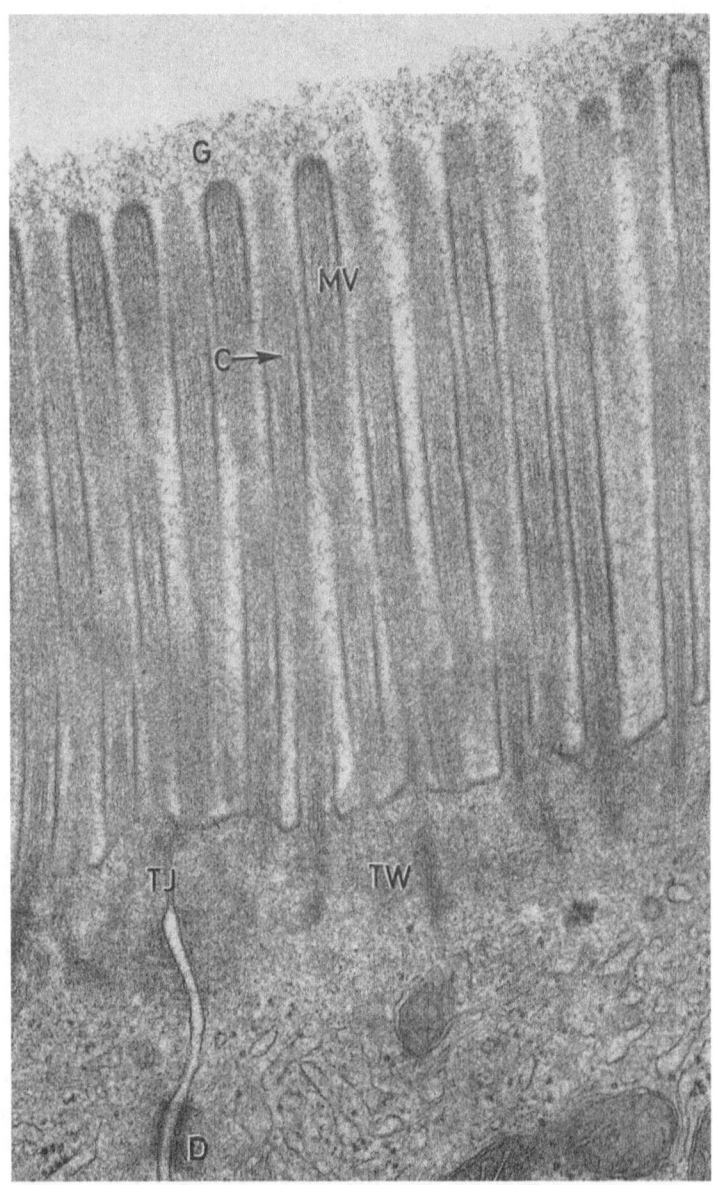

Fig. 5. Electron micrograph of the lumenal surface of human jejunal absorptive cell, showing the surface "fuzz" coat (glycocalyx, G), microvilli (MV), and the fibrils of the microvillus core (C) which enter the apical aspect of the cell and intermesh with the terminal web (TW). Also shown are tight junction (TJ) and desmosome (D) of adjacent cells. (Courtesy of Dr. ALBERT L. JONES). (× 50000)

A particularly important feature of the intestinal surface coat is the fact that it is rich in highly polar (negatively charged) sulfated mucopolysaccharides. Although this charged barrier clearly does not prevent cellular uptake of nega-

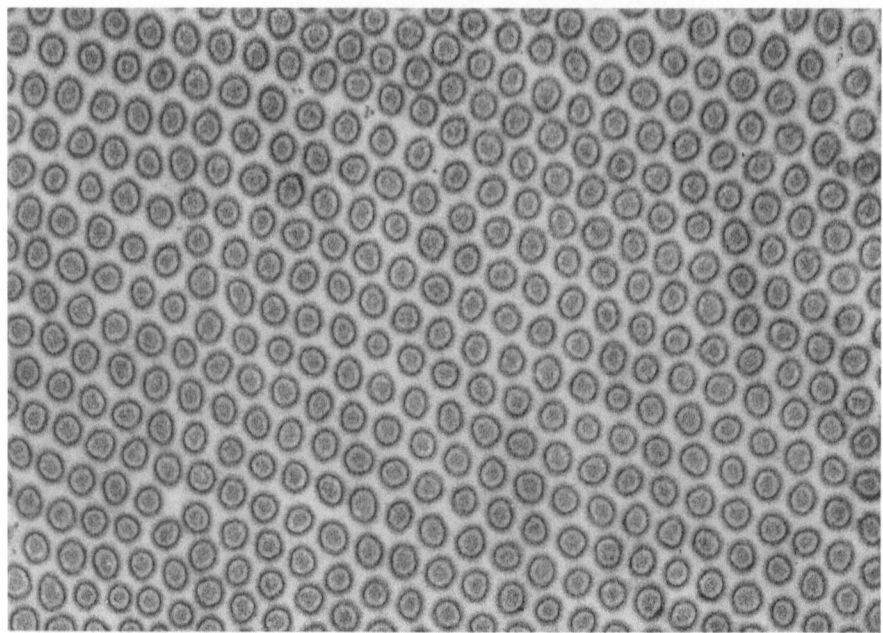

Fig. 6. Electron micrograph of rat intestinal microvilli in cross section, showing glycocalyx, plasma membrane and central core. (Courtesy of Dr. Albert L. Jones). (× 50000)

tively charged solutes, e.g. fatty acids, it may affect fatty-acid absorption in an equally profound, albeit less direct manner. For some time it has been recognized that, when an epithelial surface borders on a fluid-filled space, the portion of the fluid immediately adjacent to the cell surface may not be in equilibrium with the bulk phase and may therefore constitute a separate compartment, i.e. an "unstirred layer" (Dainty, 1963). That this unstirred layer also exists in relation to the intestinal epithelium has been re-emphasized recently by Wilson and Dietschy, who have suggested that it may be rate-limiting in the absorption of fatty acids and certain other solutes such as bile salts (Wilson et al., 1971; Wilson and Dietschy, 1972).

What is the relationship of the unstirred layer to the glycocalyx and to the fluid which exists within its interstices? A definite answer to this question is not yet available, but it has been suggested that the highly polar residues in the glycocalyx serve to "organize" the immediately adjacent water molecules into unstirred lamellae by means of an electrostatic effect (Schafer and Andreoli, 1972). Because the thickness of the unstirred layer cannot be measured directly, estimates have been based primarily on calculations derived from flux rates for solutes (with known diffusion coefficients) under various experimental conditions. It has been suggested that, for the intestine, the unstirred layer varies in overall thickness from about 250 to 500 µm, depending on the intensity with which the bulk phase is stirred (Westergaard and Dietschy, 1972). This greatly exceeds the thickness of the glycocalyx, thereby supporting the concept that it is "unstirred" rather than entrapped water. In addition, it is approximately

10 to 20 times the overall height of the absorptive cell itself, and more than 10^6 times the diameter of a water molecule.

Although these estimates of the dimensions of the unstirred layer are of some help in conceptualizing what may be occurring at the cell surface, it should be recognized that their derivation is based on a number of assumptions, not all of which can be tested directly with currently available techniques. For example, it is difficult to be certain how the normal intestine, absorbing fat *in vivo* (in the presence of peristalsis and other movements of the intestinal wall and villi, and perhaps microvilli, which ordinarily accompany this process), relates to the everted gut sacs used in the *in vitro* experiments which have been interpreted as suggesting that the unstirred layer is rate-limiting in the absorptive process. Is the intact intestine normally in an "unstirred" or a "stirred" condition? If "stirred", then, although the unstirred layer may be the rate-limiting step in the *in vitro* experimental situation, it may not, in fact, be rate-limiting *in vivo*. At the present time, it seems reasonable to conclude that placing the intestinal unstirred layer in proper physiological perspective must await further methodological advances.

B. Entry of Fatty Acid into the Microvillus Membrane

Whatever the nature of the process by which fatty acids gain access to the surface of the absorptive cell, two aspects of their uptake by the microvillus membrane seem reasonably well established. First, the mixed bile salt-monoglyceride-fatty acid micelle is not absorbed intact. Rather, the micelle dissociates, the fatty acid and monoglyceride enter the absorptive cell, and the bile salts for the most part remain within the lumen, where they participate in the solubilization of additional lipids; they are finally absorbed via an active transport system in the ileum. It is not yet entirely clear whether the bile salts transiently enter the jejunal epithelial membrane during the absorption of micellar lipids or whether they remain entirely extracellular; recent studies of WILSON and DIETSCHY (1972) favor the latter alternative.

A second aspect of long chain fatty acid absorption which seems well supported by existing evidence is that the entry of these molecules into the microvillus membrane is a passive process. Thus, fatty acid is taken up from a micellar solution at low temperatures, and in metabolically poisoned or dead tissue preparations (JOHNSTON and BORGSTRÖM, 1964). Even boiled gut sacs take up micellar fatty acids to a normal extent, suggesting that specific receptor sites (which presumably would be denatured by boiling) are not required. In these respects, uptake of luminal fatty acid by the absorptive cell seems entirely analogous to uptake by other mammalian cells, including erythrocytes (GOODMAN, 1958; SHOHET et al., 1968) and Ehrlich ascites tumor cells (SPECTOR et al., 1965). The weight of evidence suggests that the uptake process involves a hydrophobic interaction of the fatty acid with membrane lipid. This concept is further supported by the observation that fatty acids and fatty acid methyl esters are taken up equally well by ascites tumor cells (Kuhl and SPECTOR, 1970), suggesting that an intact carboxyl group is not required for the uptake process, and constituting additional evidence against the concept that specific receptor sites are involved.

The association of fatty acids with plasma membranes (intestinal, erythro-
cyte, etc.) is reversible, as might be expected of a hydrophobic lipid–lipid interaction. Thus, fatty acids, after uptake by everted intestinal sacs (Mishkin et al.,
1972), erythrocytes (Goodman, 1958), or ascites tumor cells (Spector et al., 1965),
can be removed from the cell surface and recovered in the extracellular medium.
Although such reversibility can be demonstrated to a limited extent by simply
"washing" the tissue in an isosmotic electrolyte buffer solution, removal of
fatty acids from the cell membrane is greatly enhanced if albumin is present
in the "wash" medium. This phenomenon has been observed for each of the
several tissues examined, and is not unexpected in view of the limited solubility
of long-chain fatty acids in aqueous media and their high affinity for the lipid-rich
cell membrane itself. The albumin, by serving as a soluble acceptor which also
exhibits a strong binding affinity for fatty acids, thus provides a mechanism
by which they may be removed from the membrane.

The physiological implication of these observations is of particular significance
when one considers the subsequent events, both in the absorption of fatty acids
by intestinal epithelium and in the metabolism of fatty acids by other tissues.
It is apparent that those physical properties that initially allow the cell membrane
to compete successfully with the bile-acid micelle (or with plasma albumin) and
thereby to "take up" fatty acids, could be regarded as factors which might retard
the subsequent "efflux" of fatty acids from the *inner* surface of the plasma
membrane as they undergo translocation to the aqueous interior of the cell.
It seems reasonable that some mechanism might be required to permit this
otherwise thermodynamically unlikely process (i.e. the spontaneous net flux of
long-chain fatty acids from a lipid membrane into an aqueous phase in which
they are only poorly soluble) to occur at a rate sufficiently great to account
for the rapidity with which long-chain fatty acids are activated and esterified
during the uptake or absorptive process. Indeed, recent studies have provided
evidence for the existence of such a facilitating mechanism, in the form of a
fatty acid-binding protein which is present in the cytosol of intestinal mucosa
and of other tissues which take up and utilize long-chain fatty acids (see below).

IV. Intracellular Metabolism of Absorbed Long-Chain Fatty Acids

A. Fatty Acid Binding Protein (FABP)

In studies designed to explore the mechanism underlying differences between
exogenous long-chain saturated and unsaturated fatty acids in intestinal lipoprotein formation (Ockner et al., 1969b; Ockner and Jones, 1970), it was
observed that, despite equal rates of uptake, the unsaturated fatty acid was
esterified more rapidly (Ockner et al., 1972a). Because this finding could not
be explained by differences in the specificity of activating or esterifying enzymes
(Brindley and Hübscher, 1966), it was proposed that a soluble intracellular
carrier protein with different binding affinities for various long-chain fatty acids

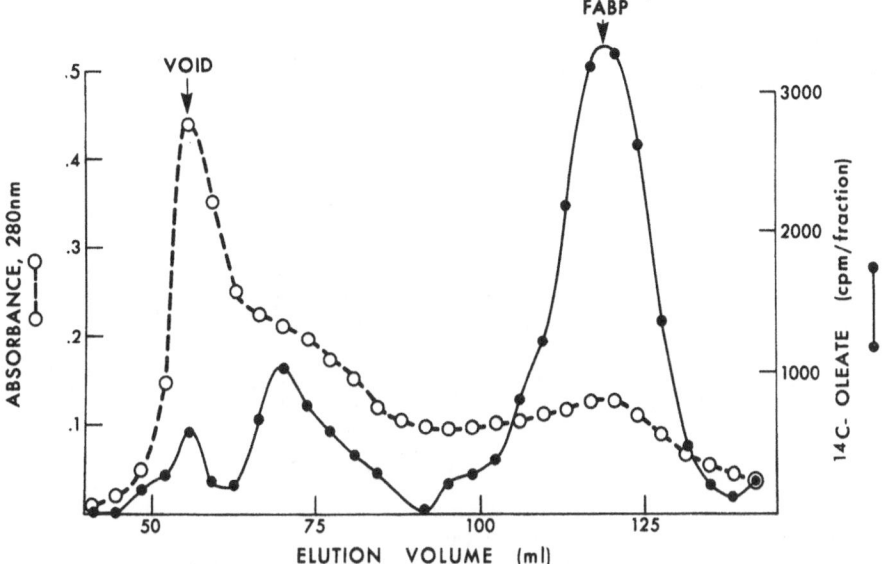

Fig. 7. Sephadex G-75 chromatography of rat jejunal 105000 g supernatant with ^{14}C-oleate. Fatty acid elutes chiefly in association with a low molecular weight fraction (elution volume ~ 120 ml) corresponding to an apparent molecular weight of 12000

might participate in the transport of fatty acids from the microvillus membrane to the smooth endoplasmic reticulum, and thereby ultimately affect esterification rates (OCKNER et al., 1972a, b). Subsequent studies (OCKNER et al., 1972b) confirmed the existence of this protein (Fig. 7), and several of its characteristics were subsequently found to be consistent with the role initially proposed for it. Thus, the protein was shown to be soluble with a low molecular weight (12000) and to have a high binding affinity for long-chain unsaturated fatty acids, lesser affinity for long-chain saturated fatty acids, and virtually no binding to medium-chain (C8 to C12) fatty acids. Binding of other lipids, including long-chain fatty acid analogs (hexadecanol and methyl palmitate), and of cholesterol was also very weak. Additional experiments showed the protein to be present in highest concentrations in the mucosa from proximal and middle thirds of rat intestine, and lowest in distal (ileal) segments; concentration of FABP in villi exceeded that in crypts. Furthermore, concentrations of FABP were significantly higher in the intestinal mucosa of rats maintained on a high-fat diet than of those on a low-fat diet (OCKNER and MANNING, 1974).

The relationship of intestinal FABP to cytoplasmic proteins in other tissues is of some interest. Several lines of evidence suggest that FABP is closely related or identical to the previously described "cytoplasmic anion binding protein" designated "Z" by LEVI et al. (1969), and initially reported as existing only in liver, intestine and kidney (Fig. 8). FABP and "Z" appear to have similar molecular weights. In addition, competition for binding between fatty acids and sulfobromophthalein (BSP, an organic anion excreted by the liver) can be demon-

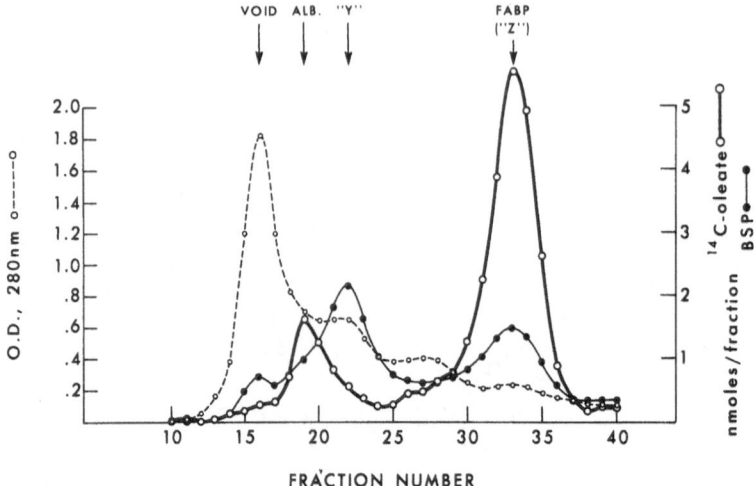

Fig. 8. Sephadex G-75 chromatography of rat liver 105000 g supernatant with ^{14}C-oleate and sulphobromophthalein (BSP). Fatty acid is eluted chiefly in association with a low molecular weight fraction (FABP) corresponding to an apparent molecular weight of 12000 and indistinguishable from the elution characteristics of the "Z" protein fraction, identified by the binding of BSP. (From Ockner et al., 1972b)

strated, further suggesting the identity of these two proteins. It is of interest, however, that FABP similar or identical to that in the intestine is present not only in liver but also in adipose tissue, myocardium, and kidney. These tissues, like the intestine, actively utilize fatty acids but do not transport BSP to a significant extent. Moreover, as determined immunochemically, the concentration of FABP in the intestinal mucosa greatly exceeds that in other tissues, including the liver, and raises serious questions regarding the role of this protein as a relatively non-specific "organic anion-binding protein".

Although FABP was initially postulated to function as a transport protein, that is, as a soluble carrier which facilitated the removal of fatty acids from the inner surface of the microvillus (plasma) membrane and participated in their translocation to the smooth endoplasmic reticulum (SER) for subsequent activation and esterification (see below), it is possible that the protein may serve other functions as well. For example, it is known that fatty acids are potentially toxic to various cellular metabolic processes, and it is conceivable that FABP, by binding free fatty acids, serves the protective function of maintaining a lower concentration of unbound FFA in cell water. Also, it may be that FABP in some manner renders long-chain fatty acids more suitable as substrates for the activating system in the smooth endpolasmic reticulum. This could reflect either the solubilization of fatty acid or an effect on the reactivity of the carboxyl group. In this regard, it is of interest that Hübscher et al. (1963) observed that the addition of the intestinal mucosal 105000 g supernatant fraction (or of an ammonium sulfate precipitable non-dialyzable fraction derived from it) to microsomal systems gretaly enhanced the conversion of long-chain fatty acid to its acyl-coenzyme A derivative. Although evidence was presented subsequently which suggested that the effect might be due to soluble phosphatidate phosphohydrolase (HÜBSCHER et al., 1967; JOHNSTON et al., 1967), a contribution by other soluble proteins was not excluded. It is entirely possible, therefore, that the enhanced fatty acid activation resulted from the presence of FABP in the preparations employed. These studies emphasize the need for further clarification of the role of FABP, not only in the intestine but also in other tissues which utilize long-chain fatty acids.

B. Fatty Acid Activation and Esterification

To this point, the events involved in fat absorption have not required direct expenditure of energy by the intestinal mucosa. Rather, by means of a series of energy-independent steps, luminal lipid has been hydrolyzed, solubilized, has entered the absorptive cell, and has gained access to the apical vesicles of the smooth endoplasmic reticulum (Fig. 9). Subsequent to this, the process requires the direct expenditure of energy by the absorptive cell for fatty acid activation, esterification, and incorporation into triglyceride-rich lipoproteins for secretion into the intestinal lymphatics. This distinction in requirements for metabolic energy between uptake and the more complex processes which follow has been documented in a series of elegant studies by STRAUSS (1968), correlating structure and function. After incubation of hamster intestine with mixed micelles at 0°, it was shown that despite continued uptake of fatty acid, esterification (reflected by incorporation of fatty acid radioactivity into triglyceride, and the appearance of lipid droplets in the endoplasmic reticulum) did not occur. If, however, the tissue was subsequently incubated at 37° in a medium containing no lipid, fatty

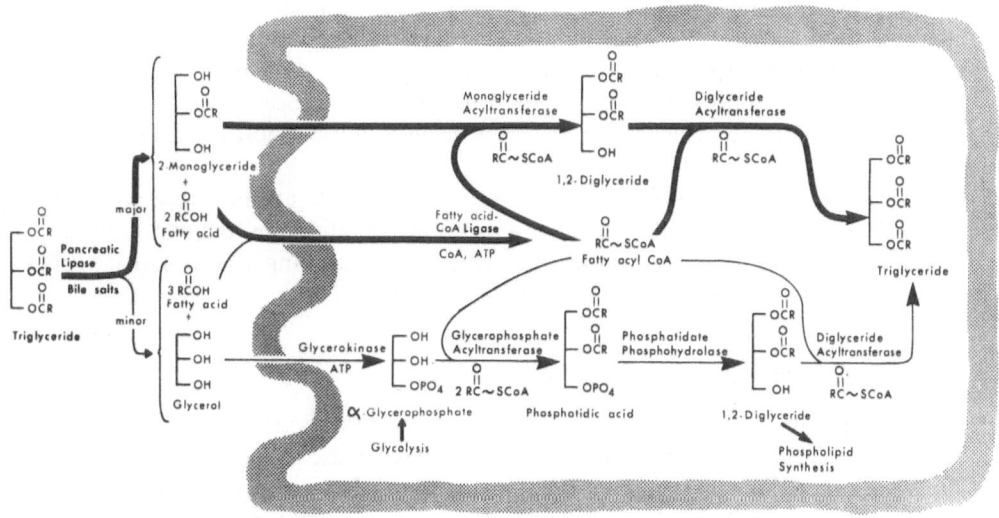

LUMEN   ABSORPTIVE CELL

Fig. 9. Pathways of triglyceride biosynthesis in the intestinal absorptive cell

acid radioactivity (previously taken up at 0° but not esterified) then was incorporated into triglyceride, and lipid droplets were demonstrable morphologically in the endoplasmic reticulum.

1. Fatty Acid Activation

As the initial energy-requiring step, activation of fatty acid to the acyl-coenzyme A derivative by microsomal fatty acid-coenzyme A ligase (thiokinase) is fundamental to subsequent metabolism and transport of absorbed lipid. In studies with microsomal fractions from cat and guinea-pig intestine, Brindley and Hübscher (1966) found that the enzyme required ATP and Mg^{++} in addition to fatty acid and coenzyme A substrates. It showed greatest activity with fatty acids of chain lengths exceeding 10–12 carbon atoms, an observation which is consistent with, and may partially explain, the fact that medium-chain and short-chain fatty acids (i.e., those with chain lengths of 12 carbons or fewer) pass through the absorptive cell and enter the portal vein without undergoing activation or esterification. The fact that some conversion of C10 and C12 fatty acids to their acyl-CoA derivatives does occur *in vitro* (Brindley and Hübscher, 1966) suggests, however, that substrate specificity of the ligase enzyme alone cannot account for the clearcut differences in absorption which have been observed between medium-chain and long-chain fatty acids (Greenberger et al., 1966).

This finding is of interest in that substrate specificity of fatty acid:CoA ligase also fails to account for certain *in vivo* and *in vitro* differences among the *long-chain* fatty acids. Thus, in everted gut sac preparations (Ockner et al., 1972a) it has been shown that, despite virtually identical uptake rates, unsaturated long-chain fatty acids are esterified with significantly greater rapidity than are saturated fatty acids (Fig. 10). Since this cannot be accounted for by differences in substrate specificity of the ligase enzyme (Brindley and Hübscher, 1966), it is explainable only by postulating either more rapid access of unsaturated fatty acid to the enzyme (in the SER) after uptake at the cell surface, or by differences in the rate of subsequent incorporation of long-chain fatty acyl esters into glycerolipids (triglyceride and phospholipid). In regard to the latter process (transacylation), however, there is no evidence for an effect of unsaturation (Johnston and Rao, 1965).

Therefore, differences in the rate at which fatty acids are translocated from the cell surface to the SER remains a likely explanation for relative rates of esterification, and is consistent with the demonstrated binding affinities of FABP for individual long-chain fatty acids (Ockner et al., 1972b).

2. Fatty Acid Esterification: Triglyceride Synthesis

The endoplasmic reticulum of the absorptive cell is also the site of incorporation of fatty acyl-CoA derivatives into more complex lipids, including triglycerides and phospholipids. Despite the fact that at least two distinct pathways each have been demonstrated for synthesis of triglycerides and phospholipids, the interrelationship and possible regulation of these pathways is poorly understood. Certain constraints in this regard *are* recognized, however. For example, in

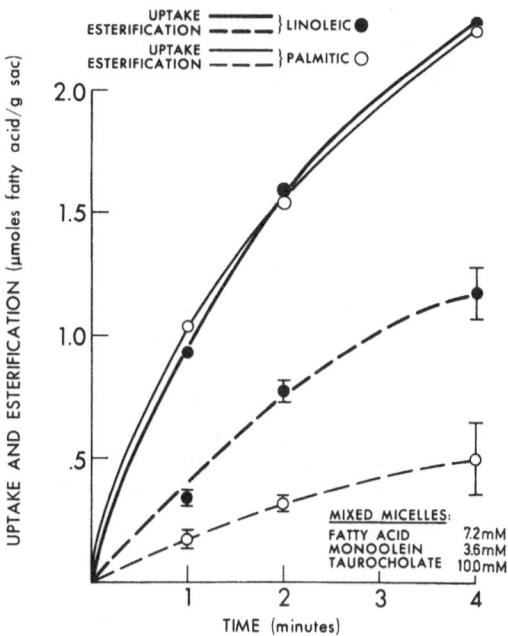

Fig. 10. Uptake and esterification of long chain unsaturated (linoleic) and saturated (palmitic) fatty acids from mixed micelles by everted gut sacs. (Modified from OCKNER et al., 1972a)

mammalian systems the 2-position of most glycerophosphatides (e.g. lecithin) is occupied by an unsaturated fatty acid; accordingly, saturated fatty acids can be expected *not* to be incorporated at this position, but aside from teleological considerations (e.g. the need to produce phospholipids with sufficiently low transition temperatures), the molecular basis for even this well-recognized example of limited specificity in fatty acid esterification is not understood. Similarly, it has been shown that cholesterol is preferentially esterified to oleic acid (KARMEN et al., 1963) but the basis for this selectivity is unknown. Thus, although the available information regarding the esterification pathways which we are about to consider has provided substantial clarification of the biochemistry of fat absorption, there is still little or no understanding of how the overall process is regulated or coordinated.

Of the two pathways for triglyceride synthesis within the intestinal absorptive cell, the first of these to be described, the α-glycerophosphate (αGP) pathway, is analogous to the pathway recognized earlier in liver (KORNBERG and PRICER, 1953). In this sequence, αGP reacts with two molecules of fatty acyl-CoA to form phosphatidic acid (Fig. 9). The phosphate ester is cleaved, forming a 1,2-diglyceride which then reacts with an additional fatty acyl-CoA to form tri-glyceride. The αGP pathway in intestinal mucosa is well established and provides a mechanism by which fatty acids, absorbed by the intestine unaccompanied by a glycerol "backbone", can be incorporated into triglyceride. The glycerol phosphate which enters this pathway is derived primarily from the glycolytic pathway (via reduction of dihydroxyacetonephosphate), but free glycerol from the intestinal lumen or plasma also may be directly phosphorylated (HAESSLER and ISSELBACHER, 1963). Ordinarily, however, because the major source of dietary fat is triglyceride, the αGP pathway appears to be of only minor quantitative significance for fat absorption and the monoglyceride pathway assumes major importance.

As has been previously discussed, luminal triglyceride is hydrolyzed primarily to fatty acids and 2-monoglyceride, both of which are incorporated into the bile salt micelle and enter the absorptive cell. Direct acylation of 2-monoglycerides, resulting in the resynthesis of triglycerides (Fig. 9), has been demonstrated by several investigators (SENIOR and ISSELBACHER, 1961, 1962; CLARK and HÜBSCHER, 1961), and subsequent studies in man have provided evidence strongly suggesting that this pathway, seemingly the most simple, direct and efficient, is also quantitatively the most important pathway utilized during the absorption of dietary fat (KAYDEN et al., 1967). Although this dominance in part reflects the fact that monoglyceride and fatty acids are the major products of intraluminal hydrolysis, JOHNSTON and his associates have recently shown that monoglyceride actually inhibits the αGP pathway of triglyceride synthesis (POLHEIM et al., 1973).

In regard to substrate specificity, studies of JOHNSTON and RAO (1965) have shown that there is no apparent preference for oleyl-CoA over palmityl CoA in the acylation of 2-monoglyceride, suggesting that saturated and unsaturated fatty acyl-CoA molecules generally are utilized to a similar extent.

In considering the two pathways for triglyceride synthesis (Fig. 9), it will be noted that in both a 1,2-diglyceride is formed as an intermediate prior to the final acylation step. Despite this, however, there is evidence suggesting that the two diglyceride pools remain distinct and do not mix (JOHNSTON et al., 1967). Moreover, in other studies it was shown that, whereas the 1,2-diglyceride formed from the αGP pathway may be converted to phosphatidyl choline, diglyceride produced via the monoglyceride pathway is not (JOHNSTON et al., 1970), thus further supporting the concept that there are, in fact, two non-miscible intracellular pools of 1,2-diglyceride.

One possible interpretation for this compartmentation is that the two pools are formed in different regions of the endoplasmic reticulum and remain associated with the specific group of enzymes involved in the pathway by which they were synthesized. This would be consistent with the concept that membrane-bound enzymes, acting in sequence in a given biosynthetic pathway, are likely to be located in close proximity to one another thereby forming a functionally distinct region of the membrane. This interpretation is strengthened by the observation of RAO and JOHNSTON (1966) that purification of an enzyme complex, designated "triglyceride synthetase", from intestinal microsomes resulted in the simultaneous purification of its component enzymes, i.e. acyl CoA:monoglyceride acyltransferase, acyl-CoA:diglyceride acyltransferase and fatty acid:CoA ligase. The authors also showed that coenzyme A itself (RAO and JOHNSTON, 1967) remained enzyme-bound.

However, any hypothesis suggesting an intracellular compartmentalization of lipid intermediates must take into account the fact that the intestinal epithelium is anatomically and functionally heterogeneous. Accordingly, what may appear

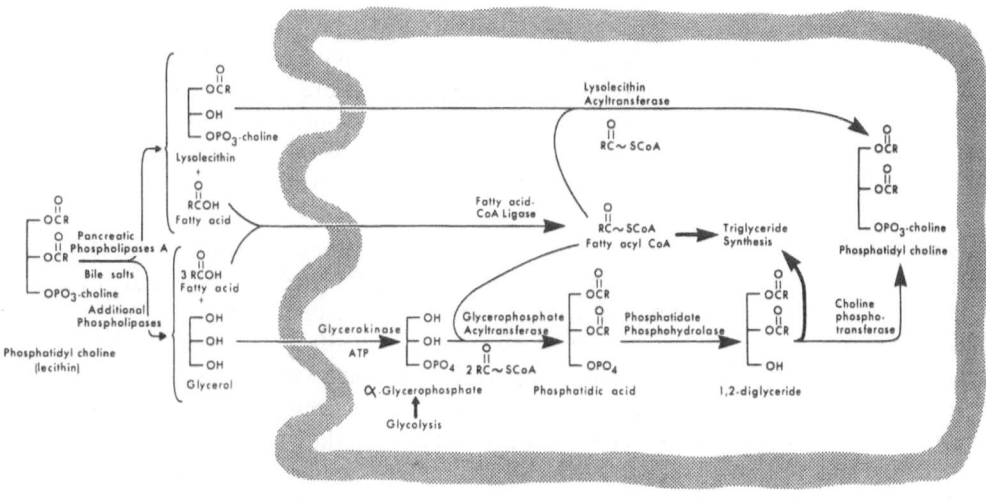

Fig. 11. Pathways of phosphatidyl choline synthesis in the intestinal absorptive cell

to be two distinct pools of a given lipid within a single cell population may in fact reflect the metabolism of that lipid by different cell types. Thus, for example, it is well recognized that intestinal epithelial cells undergo significant changes in metabolic activity from the time when they are formed in the crypts and progressively mature as they migrate up the side of the villus to the point at which they are extruded as "senescent" from the villus tip.

Recently, MANSBACH (1973) presented evidence that reflects this process in regard to lipid metabolism and may in part explain the apparent compartmentation of diglyceride pools. MANSBACH has found that in microsomes from *villus tips* the synthesis of lecithin via the direct reacylation of absorbed lysolecithin, formed in the lumen by the action of pancreatic phospholipase A and shown by SCOW (1967) to be absorbed to some extent without undergoing further hydrolysis, is more active than via the αGP pathway (Fig. 11). Furthermore, diglyceride is preferentially converted to triglyceride (rather than to phospholipid) by similar preparations from villus tips. In microsomal preparations from *crypt* cells, in contrast, lecithin synthesis appears to depend equally on the acylation of lysolecithin and on the activity of cholinephosphotransferase, by which diglyceride (formed by the αGP pathway) is converted to lecithin by the addition of phosphorylcholine.

Taken together, these observations suggest that in *villus tips*, triglyceride and lecithin are synthesized primarily by the direct acylation of their *partial* hydrolysis products (2-monoglyceride and lysolecithin, respectively) formed in the lumen by the action of appropriate pancreatic lipases and absorbed intact. Most of the diglyceride synthesized in this region is produced, therefore, by the acylation of monoglyceride rather than as an intermediate in the αGP pathway of triglyceride or lecithin synthesis. In the *crypt cells*, in contrast, less of the luminal lipid is available, but there is a need for membrane phospholipid synthesis by these rapidly proliferating and developing cells. Here, therefore, the more general αGP pathway assumes importance and provides a mechanism for the synthesis of lecithin (and other glycerolipids) from simple precursors, including fatty acids, which are derived in part from plasma (GANGL and OCKNER, 1974) and glucose.

Although this analysis must be regarded as tentative, it provides a reasonable explanation by which existing data can be reconciled and interpreted, and takes into consideration current understanding of metabolic needs and substrate availability in different anatomic regions of the absorptive epithelium.

The concept that availability of substrate may affect enzymes involved in mucosal lipid synthesis is further supported by the substantial body of evidence which indicates that the intestinal mucosa exhibits adaptive changes in response to increases in dietary fat intake. Under ordinary conditions, most ingested fat is absorbed by the proximal portions of the intestine, i.e. the jejunum (SENIOR, 1964). If dietary fat intake increases (HOVING and VALKEMA, 1969; SINGH et al., 1972), however, or if a portion of the jejunum is bypassed or removed (RODGERS and BOCHENEK, 1970) the more distal (ileal) portions of the intestine, which normally absorb only small amounts of fat, adapt to this increased "substrate" load and exhibit an enhanced absorptive capacity. The studies of RODGERS and his associates (RODGERS and BOCHENEK, 1970; SINGH et al., 1972) show that this enhancement results, not only from an increase in villus height and number of absorptive cells (i.e. hyperplasia) (DOWLING, 1973), but also from an increase in the specific activity of certain key enzymes involved in the synthesis of triglyceride via the monoglyceride pathway, including fatty acid-coenzyme A ligase and monoglyceride acyltransferase. Conversely, if the availability of lipid substrate is reduced (e.g. in bile diversion), then the activity of these enzymes is diminished. The mechanism by which these changes are induced is not known. Whether they truly represent examples of "substrate induction" or a less direct process, possibly hormonally mediated, has not been resolved.

Intestinal triglyceride synthesis also may be modulated by a possible intracellular effect of bile salts. Although bile salts are absorbed primarily in the terminal ileum, a small amount is absorbed by passive diffusion into the jejunum and measurable bile-salt concentrations have been recorded in jejunal mucosa (HOLT, 1964). *In vitro* evidence suggested that bile salts enhanced the incorporation of fatty acids into triglyceride by mucosal preparations (DAWSON and ISSELBACHER, 1960), and this finding was supported by later evidence that, in the absence of bile salts, absorbed fatty acids may be transported unesterified via the portal vein (GALLAGHER et al., 1965). Although this concept was subsequently challenged (DIETSCHY, 1967), more recent *in vivo* studies provide supportive evidence that conjugated bile salts in some manner enhance the esterification of long-chain fatty acids in the absorptive cell (CLARK et al., 1969). The mechanism for this effect is not known, but probably relates to the surface-active properties of the bile salt on the substrate, the membrane-bound enzymes involved, or both. It should be noted, however, that intracellular bile-salt concentrations are below the CMC (HOLT, 1964), and therefore micelle formation probably is not involved in the effect.

V. Intestinal Lipoprotein Production

After triglyceride, phospholipid and cholesterol ester have been synthesized by the absorptive cell, these lipids are incorporated into lipoprotein particles (including chylomicrons) which are secreted by the cell and enter the lacteals, intestinal lymphatics, thoracic duct, and finally the plasma. The mechanism by which the lipoprotein particle is assembled is not understood. However, the electron microscope has shed considerable light on the morphological aspects of this process.

The studies of CARDELL et al. (1967) confirm earlier findings of PALAY and KARLIN (1959), and show clearly that lipid droplets (triglyceride) which appear initially within the smooth endoplasmic reticulum, subsequently accumulate in the Golgi apparatus and are discharged into the lateral intercellular spaces (Fig. 12). The precise role of the Golgi apparatus in this process has not been fully defined. It is presumed that, in a manner analogous to its function in certain exocrine glands (e.g. pancreas), it somehow participates in the "packaging" of the material to be exported (the lipoprotein particle), perhaps including the investment of lipids with a polar, protein-containing coat (see below). Alternatively, its function may be more limited, and relate to the incorporation of glycoprotein into the lipoprotein particle, a possibility suggested by observations that lipoproteins contain glycoproteins, and that the glycosyltransferases of the absorptive cell are located primarily in the Golgi apparatus (LO and MARSH, 1970). Despite these considerations, however, it is not entirely clear that passage through the Golgi apparatus is an obligatory step in the secretion of lipoproteins (JONES et al., 1966, 1967).

Despite uncertainties as to the mechanism, the time-sequence studies of JERSILD (1966) have emphasized the rapidity with which the overall process occurs. These experiments involved the intrajejunal instillation of fatty chyme, followed by electron microscopic examination of the absorptive cells at timed intervals. It was demonstrated that the entire process of fat absorption including appearance of lipid droplets in the apical vesicles of the endoplasmic reticulum, accumulation of particles in the Golgi, and the secretion of lipoproteins into the intercellular space, was proceeding maximally within 12 minutes. There must be few metabolic processes which are at once so complex and yet so rapidly responsive to abrupt increases in substrate load as is the intestinal absorption of fat.

The lipoproteins formed by the intestine historically have been designated "chylomicrons", a term originally used by GAGE and FISH (1924) to describe particles which appeared in blood plasma after a fatty meal and which were visible microscopically. It is now known that these relatively large particles have a diameter in excess of 1000 Å and are composed predominantly of triglyceride, with lesser amounts of phospholipid, free and ester cholesterol, and protein. In studies designed to determine the structure and composition of lymph chylomicrons, ZILVERSMIT concluded that the nonpolar lipids (triglyceride and cholesterol ester) were contained within the core of the particle and were surrounded by a more hydrophilic "membrane" consisting of protein, phospholipid and free cholesterol (ZILVERSMIT, 1968; SALPETER and ZILVERSMIT, 1968). Although

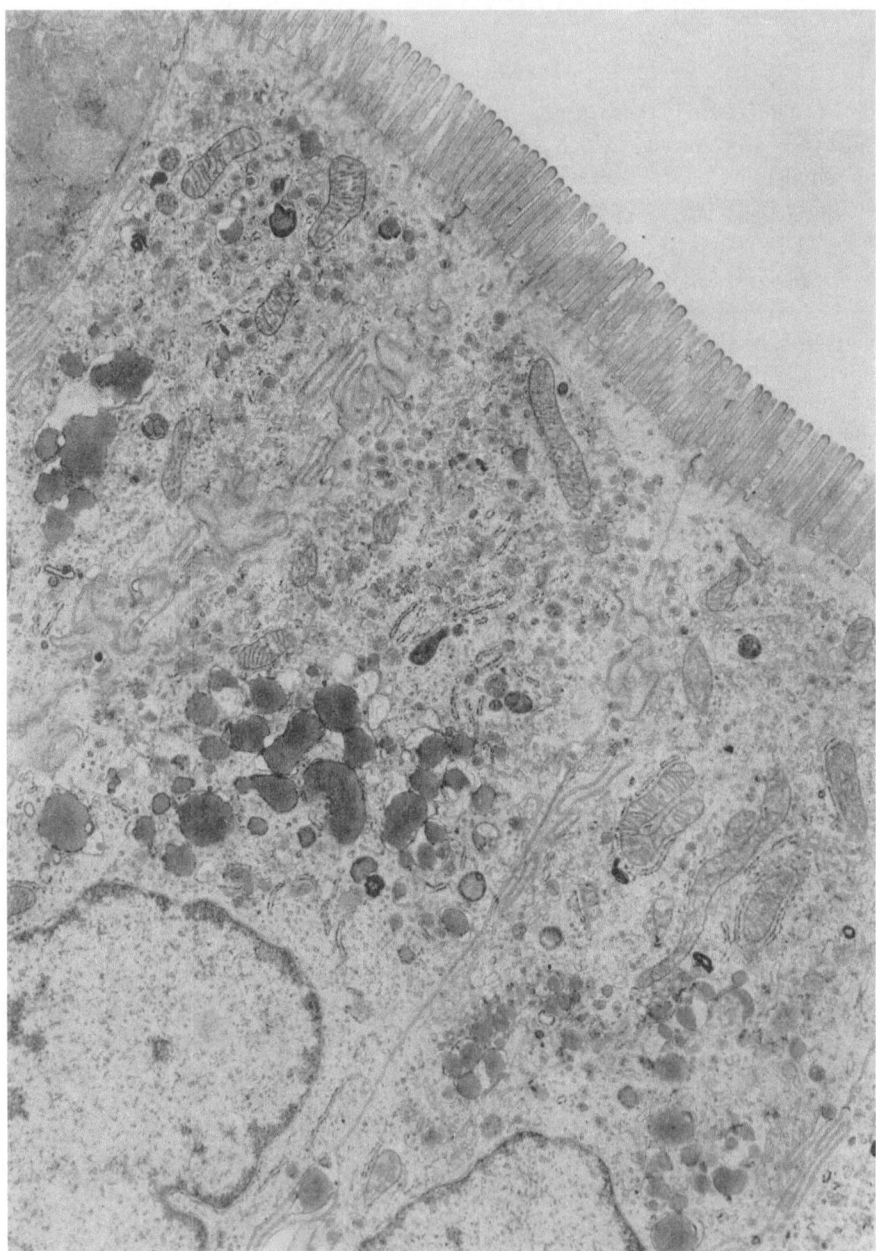

Fig. 12a. Electron micrograph of the intestinal absorptive cell during fat absorption. See Fig. 12b. (From Cardell et al., 1967. (× 12000)

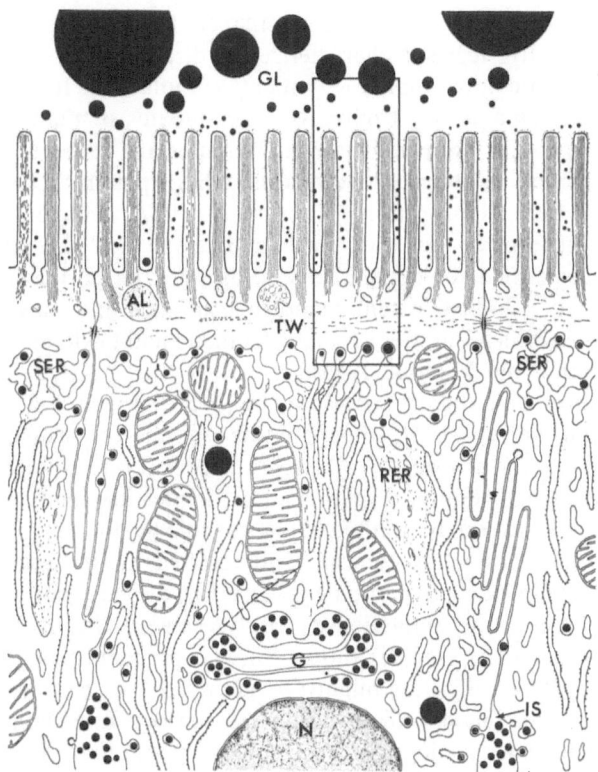

Fig. 12b. Diagrammatic representation of the intestinal absorptive cell during fat absorption showing gut lumen (GL), electron-dense (osmiophilic) particles in the smooth endoplasmic reticulum (SER), Golgi apparatus (G), and intercellular spaces (IS). Also depicted are the terminal web (TW), apical lysosomes (AL), rough endoplasmic reticulum (RER), and nucleus (N). (From CARDELL et al., 1967)

it now seems clear that a true "unit" membrane is not present, recent studies of other lipoprotein systems support ZILVERSMIT's concept of the spatial orientation of the components of the lipoprotein particle, which thus may be regarded as analogous to the micelle in that the polar components (capable of a stable interaction with water) are exterior to a hydrophobic "core" (STEIM et al., 1968).

A. Lipoprotein Peptides

The nature and function of the lipoprotein peptides has attracted considerable interest recently. In regard to the intestine, the importance of chylomicron protein was highlighted by observations of patients with the rare disorder abetalipoproteinemia. This disease, recently shown to be associated with undetectable levels of normal apo-LDL (or "B protein"), is characterized by the absence of low-density lipoproteins from plasma and, accordingly, extremely low plasma cholesterol levels (Salt et al., 1960; Isselbacher et al., 1964; Gotto et al., 1971). Several other distinctive features characterize the disorder, but it is of particular interest in the present context that these patients are unable to form chylomicrons after a fatty meal; as a result, triglyceride accumulates in the absorptive cells and fat malabsorption is present. It is equally important to note, however, that despite their inability to form chylomicrons, these patients characteristically exhibit only *mild* fat malabsorption (Ways, et al. 1967), implying the existence of an alternate absorptive pathway. Such a pathway, i.e. the direct entry of absorbed but unesterified long-chain fatty acids into the portal vein, has been shown to occur to a limited extent in normal animals and may be increased under conditions in which intestinal protein synthesis and chylomicron formation is inhibited (Kayden and Medick, 1969).

The observation that these patients are unable to form chylomicrons gave rise to the concept that apo-LDL somehow was required for the exit of the lipoprotein particle from the absorptive cell. It was reasoned, therefore, that a similar circumstance might be produced experimentally if intestinal synthesis of protein (including apo-LDL) could be inhibited pharmacologically. Initial experiments with puromycin supported this concept, and resulted in the failure of rats so treated to exhibit alimentary hyperlipidemia, despite the fact that lipid had been absorbed, as shown by accumulation of triglyceride in the intestinal mucosa (Sabesin and Isselbacher, 1965). Although these experiments suggested that the use of protein synthesis inhibitors might provide a valid model for the intestinal defect in abetalipoproteinemia, this concept was subsequently challenged on the grounds that such pharmacological agents might have other effects (including delayed gastric emptying) that could complicate the interpretation of the experimental results (Redgrave and Zilversmit, 1969). The matter remained largely unresolved until the recent studies of Glickman et al. (1972), designed to circumvent the above problems, showed clearly that under conditions of inhibited protein synthesis there was diminished recovery of labeled fatty acid from intestinal lymph after intraduodenal administration of micellar lipid, and that the lymph chylomicrons under these conditions showed an increase in size, i.e. a decrease in surface-to-volume ratio. These experiments showed that inhibition of protein synthesis did indeed impair fat absorption, and in addition led to the formation of chylomicrons which presumably had a lower protein:triglyceride ratio, a phenomenon which could be regarded teleologically as a means by which lipoprotein peptides might be conserved when in short supply. In an extension of this work (Glickman and Kirsch, 1973), it was shown that cycloheximide had a particularly pronounced affect on a specific chylomicron peptide, which was immunologically related to plasma high-density lipoproteins. Although these studies do not directly link inhibition of protein synthesis to abetalipoproteinemia, they clearly indicate that this model is associated with abnormalities in the intestinal lipoprotein secretion, and further support the concept that protein synthesis plays a critical role in this process.

Chylomicrons and the other triglyceride-rich lipoproteins formed in the intestine contain peptides other than apo-LDL. Indeed, the more relatively recently recognized group of low molecular weight "C" peptides (also referred to as "D" peptides by later investigators) have been shown to account for a significant portion of the total protein present in both lymph and plasma very-low-density lipoprotein (VLDL) (BROWN et al., 1969), and were first described in particles isolated from human chyle (ALAUPOVIC et al., 1968). It has become evident that certain of these apoproteins play an important role in the activation of lipoprotein lipase, thereby facilitating the clearance of chylomicron and VLDL triglyceride from plasma (BIER and HAVEL, 1970). It is of interest, however, that despite the presence of these peptides in particles recovered from intestinal lymph, they do not seem to be synthesized in the intestine but rather in the liver (WINDMUELLER and SPAETH, 1972). Recent studies suggest that they are readily exchangeable, and that as triglyceride is removed from chylomicrons or VLDL in peripheral tissues, the small peptides transfer to high-density lipoproteins (HDL), which then constitute a continuously available source of the peptides for the surface of the newly secreted intestinal particles (HAVEL et al., 1973).

Thus, proteins play a crucial role in at least two aspects of the lipoprotein transport of lipid absorbed from the intestine. Apo-LDL appears to be required for secretion of the lipoprotein particle out of the cell and into lymph. The small peptides, on the other hand, associate with the particle only after it enters lymph and plasma, and are necessary for the activation of lipoprotein lipase in the capillary endothelium, thereby permitting the entry of fatty acids and 2-monoglyceride into adipose and other peripheral tissues.

B. Endogenous Intestinal Lipoproteins

Thus far in our discussion of the intestinal absorption of fat, it has been implicit that the lipid which undergoes luminal hydrolysis, uptake, esterification and lipoprotein transport is dietary in origin. Indeed, most of the studies which have contributed to the elucidation of these pathways have emphasized exogenous lipid primarily. The studies of several groups of investigators, however, leave no doubt that, even in the absence of dietary fat, the intestine is continuously involved in the process of absorbing and transporting *endogenous* lipid and of secreting endogenous lipoproteins into lymph (BAXTER, 1966; ROHEIM et al., 1966; OCKNER et al., 1969a; OCKNER and JONES, 1970; WINDMUELLER et al., 1970; TYTGAT et al., 1971). Moreover, during the absorption of dietary lipid as well, it is clear that a significant portion of the lipid incorporated into the chylomicron is *endogenous* in origin (KARMEN et al., 1963). In order to understand these broader aspects of intestinal lipid transport and their physiological implications, it is necessary to consider the sources of endogenous lipid available to the intestine, and the nature of the endogenous lipoprotein particles secreted by the intestine into lymph.

Endogenous lipids enter the intestinal lumen either from bile, which is rich in phospholipid (chiefly lecithin) and cholesterol, or from the gastrointestinal mucosa as it is shed during the process of epithelial cell turnover. These lipids are hydrolyzed and for the most part reabsorbed. Additional non-luminal lipid is available to the absorptive cell from two sources including *de novo* fatty acid synthesis in the intestinal cell itself (FRANKS et al., 1966), and plasma free fatty acids (FFA) (GANGL and OCKNER, 1974). The net result is a continuous (but probably fluctuating) supply of lipid which enters the absorptive epithelium and

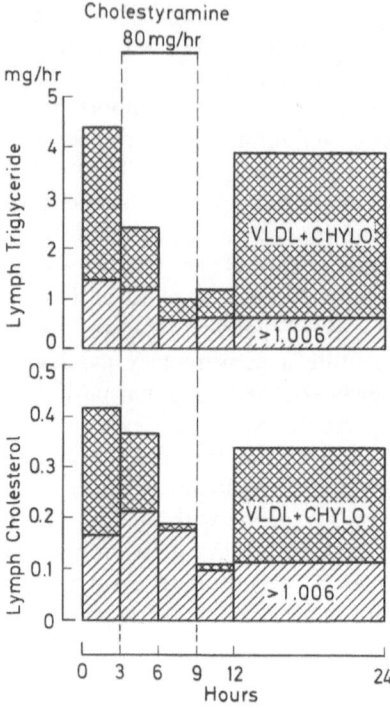

Fig. 13. Effect of introduadenal cholestyramine infusion on lipids and lipoproteins in fasting rat intestinal lymph. (From Ockner et al., 1969)

s handled in a manner analogous to dietary lipid. Thus, the intestinal lymph of a fasting rat, or one on a fat-free intake, will contain triglyceride-rich lipo-proteins. If under these conditions the supply of endogenous luminal lipid is reduced or eliminated, e.g. by means of bile diversion or administration of a bile-salt sequestrant (cholestyramine resin), the output of lymph triglyceride-rich lipoproteins will virtually cease (Ockner et al., 1969b; Ockner and Jones, 1970) (Fig. 13). These observations indicate not only that these endogenous lymph lipoproteins are formed from luminal lipid (and therefore are produced in the intestine itself), but also that the non-luminal sources (*de novo* fatty acid synthesis, or uptake of plasma FFA) are ordinarily of minor quantitative significance.

Other studies, employing the electron microscope, have provided additional confirmation of endogenous intestinal lipoprotein production. Thus, in both man and rat, lipoprotein particles similar to those found in the isolated perfused liver during lipoprotein production (Jones et al., 1967) are visible in the endo-plasmic reticulum (Fig. 14) and Golgi apparatus (Fig. 15) of the jejunal absorptive cell, and within the intercellular spaces and lacteals (Jones and Ockner, 1971). Within 6 hours of bile diversion or cholestyramine administration in the rat, these particles almost completely disappear (Fig. 16).

It seems clear, therefore, that the non-dietary lipoprotein particles present in intestinal lymph are produced and secreted by the intestine itself, and are

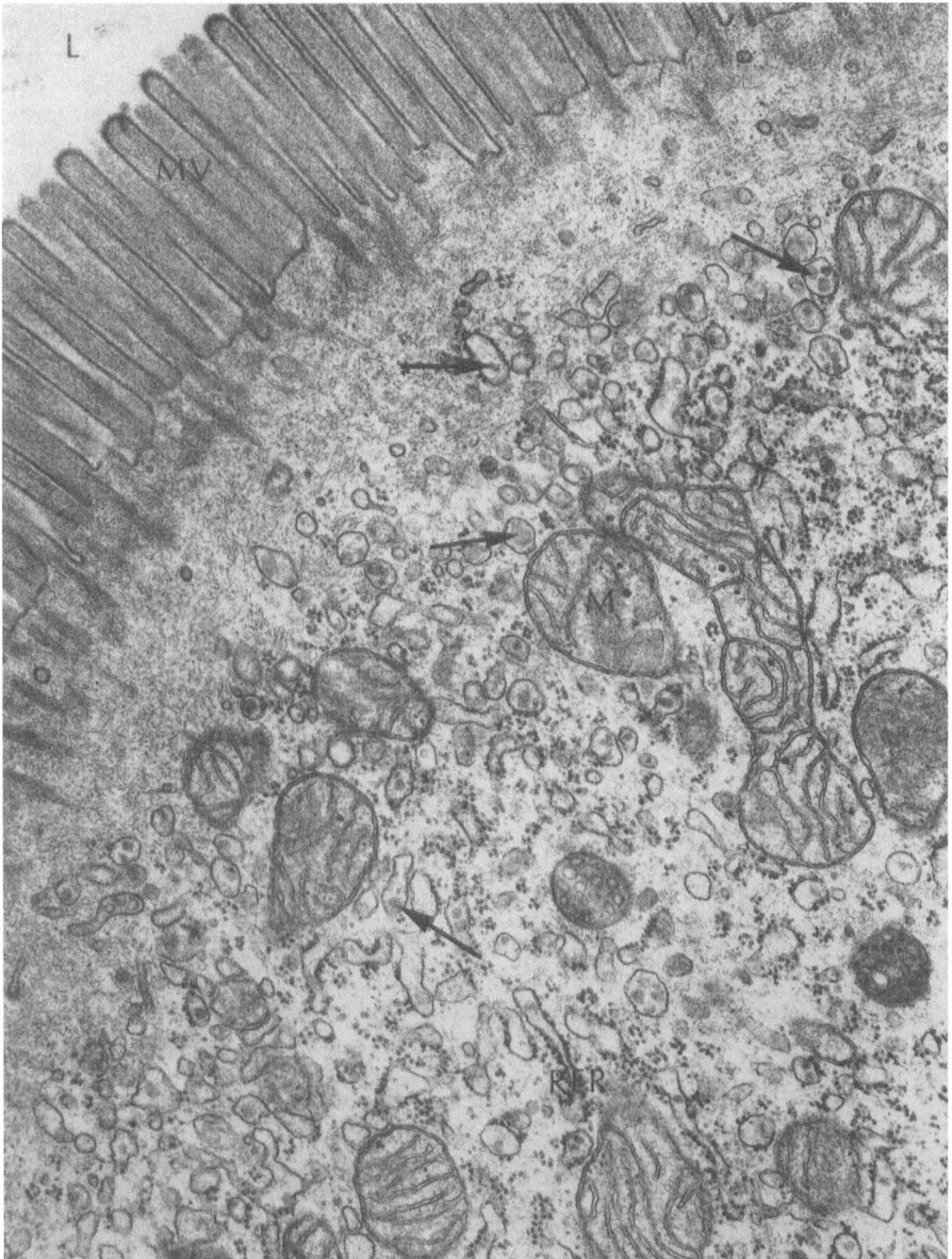

Fig. 14. Electron micrograph of apical cytoplasm of jejunal absorptive cell from a fasted restrained rat, showing osmiophilic particles within vesicles of the smooth endoplasmic reticulum
(arrows) (L, lumen; MV, microvilli; M, mitochondria; RER, rough endoplasmic reticulum).
(From JONES and OCKNER, 1971). (× 30000)

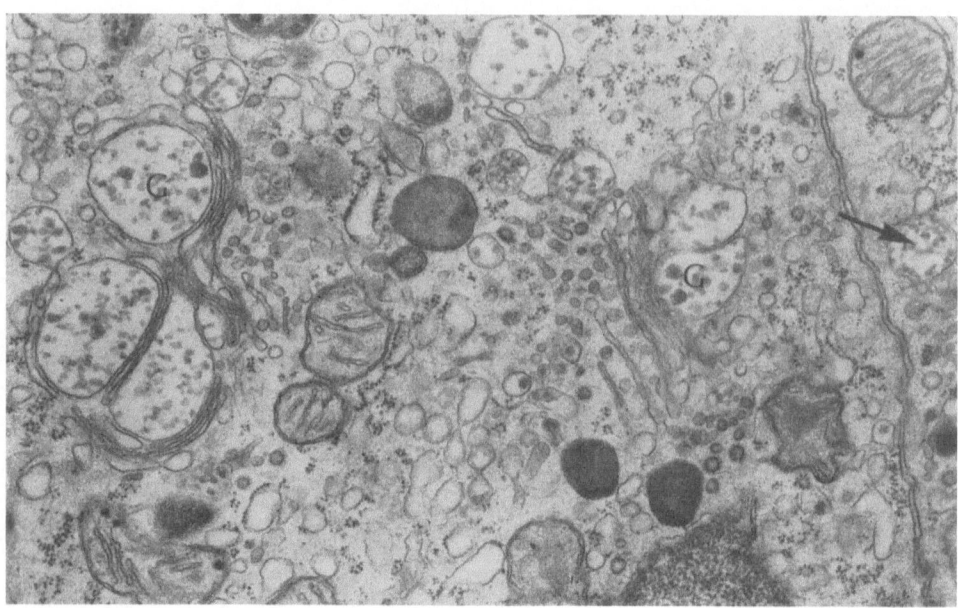

Fig. 15. Electron micrograph of Golgi (G) region of intestinal absorptive cell from a fasted restrained rat, showing osmiophilic particles within the Golgi vesicles. (From Jones and Ockner, 1971) (× 25000)

not derived from plasma by movement from the mesenteric blood capillary bed into the lymphatics. The physiological significance of this phenomenon, aside from the fact that it provides a mechanism by which endogenous luminal lipids may be reabsorbed (and thus reutilized), lies in the nature of the endogenous intestinal lymph lipoprotein particles themselves. In contrast to the relatively large chylomicrons traditionally associated with the transport of absorbed dietary fat, the endogenous intestinal lipoproteins are smaller and conform in several important respects to the properties characteristic of plasma VLDL. Morphologically they are indistinguishable (Fig. 17). Furthermore, they move with an alpha-2 ("pre-beta") electrophoretic mobility on agarose gel (chylomicrons remain at the origin), exhibit a flotation constant of S_f 20–400 (chylomicrons are $S_f > 400$) and have an overall lipid and protein composition similar to that of plasma VLDL, reflecting the presence of relatively greater quantities of surface components (protein and phospholipid), as might be expected of smaller particles that have a greater surface-to-volume ratio (Ockner et al., 1969a).

The implication of these similarities to plasma VLDL (which are generally regarded as being of hepatic origin) is that the intestine must also be regarded as a source of endogenous plasma VLDL, since they are not distinguishable by ordinary methods (such as flotation or electrophoresis) from those originating in the liver.

It must be acknowledged, however, that intestinal and hepatic VLDL do differ, at least initially, in regard to peptide composition. First, it has been shown that intestinal lymph VLDL, although they migrate electrophoretically to the pre-beta region, nevertheless are slower than plasma VLDL. If, however, lymph VLDL

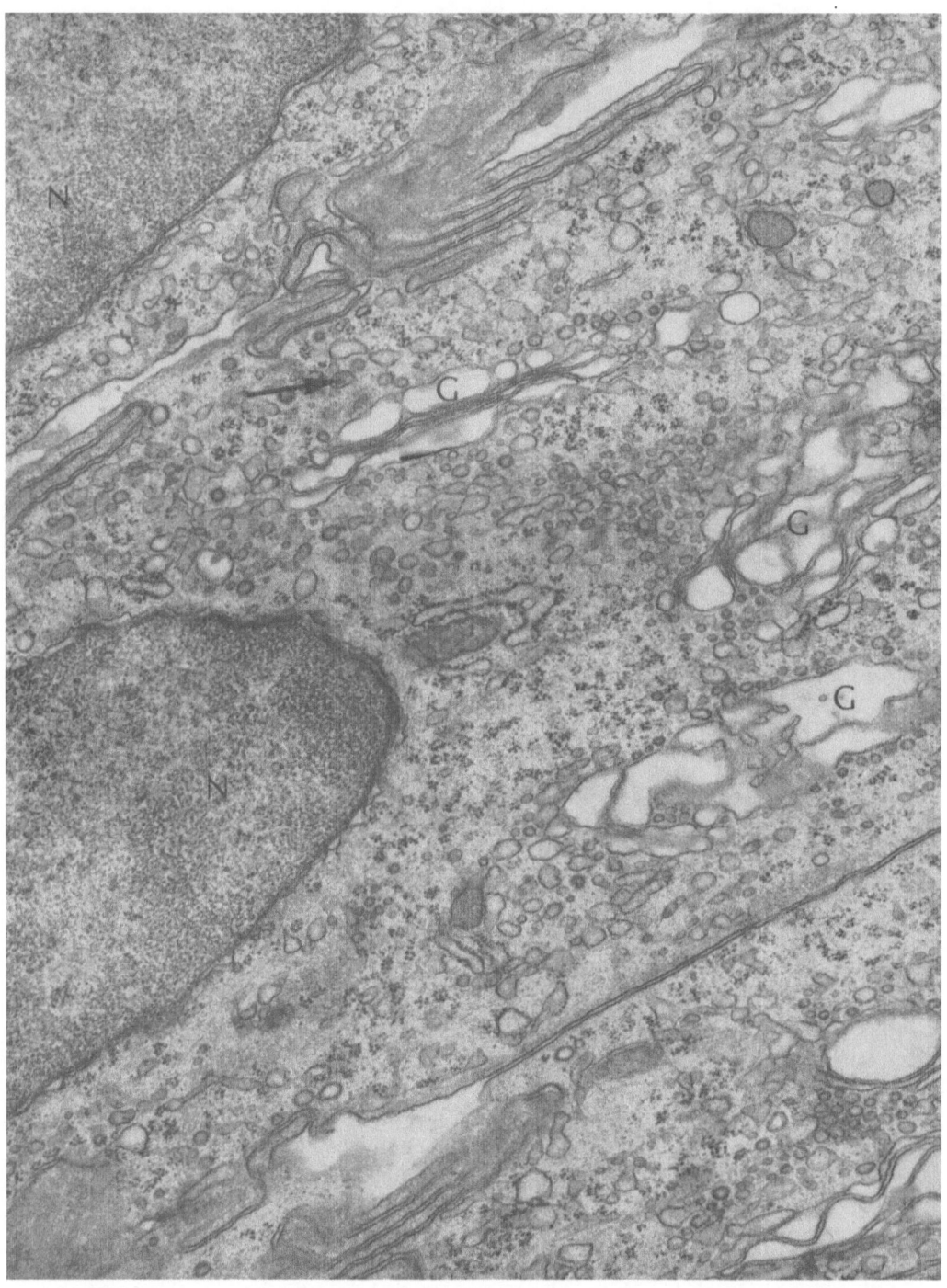

Fig. 16. Electron micrograph of supranuclear (N, nucleus) portions of intestinal absorptive cells from a fasted restrained rat, 6 hours after diversion of bile, showing virtual disappearance of osmiophilic lipid particles from the Golgi vesicles (G) and intercellular spaces. (From JONES and OCKNER, 1971). (× 23000)

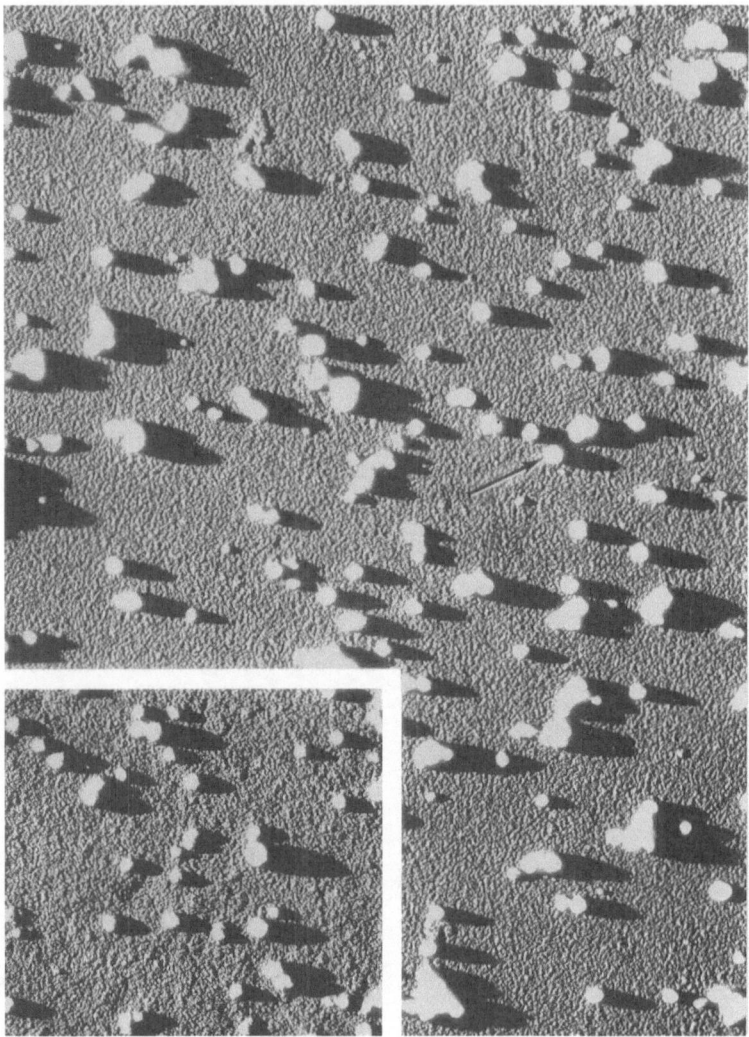

Fig. 17. Comparison of shadow-casted VLDL from intestinal lymph and plasma (inset) of fasted rat, showing similarity of particle size distribution. Particle indicated by arrow is 700 Å in diameter. (From Ockner and Jones, 1970)

are first mixed with VLDL-free serum and then immediately subjected to electrophoresis, their mobility is identical to that of plasma VLDL (Fig. 18), implying that their surface charge has increased, presumably as a result of the adsorption of additional peptides (Ockner et al., 1969a). Consistent with this is the observation alluded to above, that although intestinal lipoproteins contain the small ("activator") peptides, these do not appear to be synthesized in the intestine itself. Rather, after secretion from the intestinal absorptive cell into lymph and plasma, the surface of the lipoprotein particle acquires these additional peptide

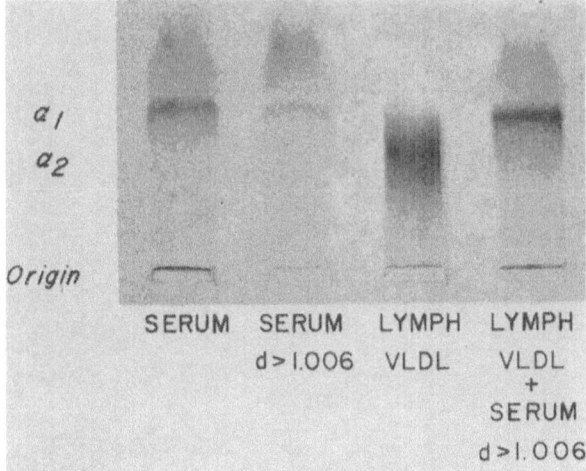

Fig. 18. Agarose gel electrophoresis of serum and lymph lipoproteins from fasting rat. On the extreme right, lymph VLDL (D< 1.006) were mixed with the serum D> 1.006 fraction immediately prior to electrophoresis. (From OCKNER et al., 1969)

components. The net result is the circulation in plasma of an intestinal particle that is metabolized by the same mechanism as hepatic VLDL (BRUNZELL et al., 1973).

The *quantitative* significance of this intestinal source of plasma VLDL remains uncertain. Although an approximation can be made that, for the rat, between 20 and 40% of the total plasma VLDL turnover is attributable to the intestine (OCKNER et al., 1969a), this estimate requires comparisons among data obtained under widely differing experimental conditions. For man, estimates are even more hazardous, although it is worth nothing that the calculated normal rate of plasma VLDL triglyceride turnover in man (about 12 g/day) (HAVEL, 1969) is similar to the estimated rate of entry of biliary phospholipid into the intestinal lumen (about 10 g/day) (GRUNDY et al., 1970) and that additional amounts of lipid enter the lumen from shed epithelium. In man, therefore, it seems likely that a significant fraction of endogenous plasma VLDL may be of intestinal origin. The potential importance of this source in certain hyperlipidemic states is supported by the observation that ethanol administration, which is associated with an increase in intestinal mucosal triglyceride synthesis (CARTER et al., 1971) and concentration, also leads to an increased rate of endogenous VLDL production by the intestine (MISTILIS and OCKNER, 1972). It is possible, therefore, that non-dietary intestinal lipoproteins contribute to ethanol-induced hyperlipidemia and conceivably to other hyperlipidemic states as well (MISTILIS and OCKNER, 1972; DEN BESTEN et al., 1973).

Other studies provide evidence that the intestine is a source of plasma VLDL not only in the fasting state but also during the absorption of exogenous fat. Thus, after intraduodenal infusion of mixed micelles, the nature of the lipoprotein particles which are secreted into intestinal lymph depends on whether the micellar

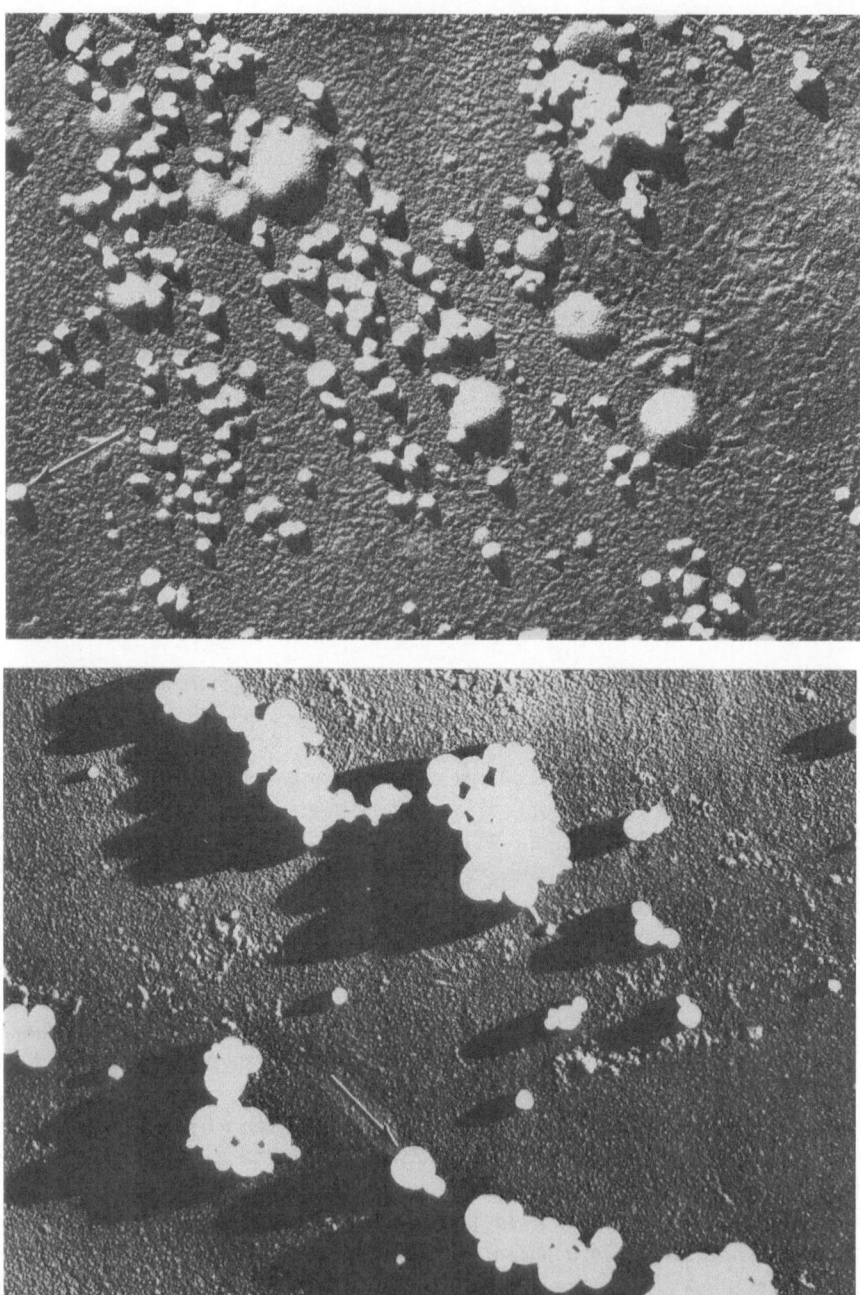

Fig. 19. Electron micrograph of shadow-casted intestinal lymph lipoproteins after intraduodenal administration of mixed micelles containing saturated (palmitic, above) and unsaturated (linoleic, below) long chain fatty acids, showing differences in particle size and distribution. The particles which appear to be very large in the upper figure actually are collapsed and flattened lipoproteins, reflecting the poor fixation of these relatively saturated particles. Particles at arrows: 700 Å (top); 1 800 Å (bottom). (From OCKNER and JONES, 1970)

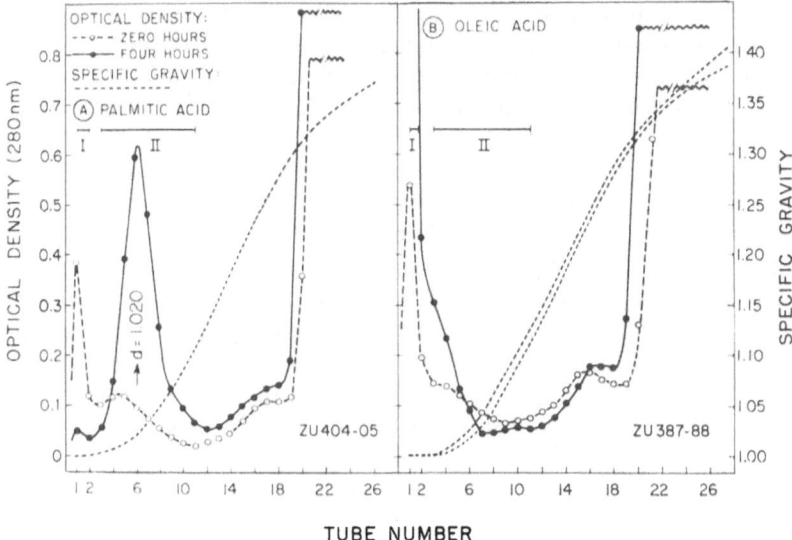

Fig. 20. Zonal ultracentrifugation analysis of lipoproteins secreted by isolated liver, perfused with medium to which either palmitic or oleic acid had been added. The rapid flotation of particles produced with oleate (peak appearance in tubes 1 and 2) contrasts with the delayed flotation of those produced with palmitate, and reflects the lower density (larger size) of the former. (From HEIMBERG and WILCOX, 1972)

fatty acid is saturated or unsaturated. With linoleate (unsaturated), lymph particles are of a size and density range typical for chylomicrons (> 800 Å, $S_f > 400$), whereas with the saturated palmitic acid, lymph particles resemble VLDL (OCKNER et al., 1969b; OCKNER and JONES, 1970) (Fig. 19). HEIMBERG and WILCOX (1972), in later studies of lipoprotein formation by the isolated perfused liver, obtained virtually identical results (Fig. 20). Together, these findings indicate that fatty acid composition is an important determinant of the physical properties of triglyceride-rich lipoproteins secreted by both intestine and liver. In the special case of the intestine, the implication is that VLDL, produced both in the absence of dietary fat and during the absorption of saturated fat, enter plasma and contribute to measured plasma VLDL and lipid levels.

VI. Conclusions

It is apparent that, despite the substantial progress of the past few years, there remain many important challenges to investigators in the area of intestinal fat absorption and transport. Perhaps foremost among these challenges is the further delineation of the relationship of intestinal lipoprotein production to systemic lipoprotein metabolism. The importance of this area is underscored not only by recent evidence that the intestinal production of lipoproteins can be modulated qualitatively and quantitatively by dietary and pharmacological means, but also by the persisting uncertainty surrounding the etiology of many of the clinically significant hyperlipidemias. An equally challenging and potentially highly rewarding area for exploration is the manner in which the intestinal absorptive cell metabolizes fatty acids. It should be recalled that this cell population is unique in that it utilizes fatty acids derived from two different sources (intestinal lumen and plasma), and preliminary evidence (GANGL and OCKNER, 1974) indicates that fatty acids from these two sources are metabolized quite differently. Elucidation of the basis for this apparent intracellular compartmentation should help to clarify other aspects of the regulation of cellular fatty acid metabolism in general. Similarly, identification of a fatty acid-binding protein in jejunal mucosa, and its subsequent demonstration in many tissues, is further evidence of the breadth of information to be gained from study of this phylogenetically primitive yet remarkably specialized organ for the transport of lipid nutrients.

References

ALAUPOVIC, P., FURMAN, R. H., FALOR, W. H., SULLIVAN, M. L., WALRAVEN, S. L., OLSON, A. C.: Isolation and characterization of human chyle chylomicrons and lipoproteins. Ann. N.Y. Acad. Sci. **149**, 791–807 (1968).

AMENT, M. E., SHIMODA, S. S., SAUNDERS, D. R., RUBIN, C. E.: Pathogenesis of steatorrhea in three cases of small intestinal stasis syndrome. Gastroenterology **63**, 728–747 (1972).

BAXTER, J. H.: Origin and characteristics of endogenous lipid in thoracic duct lymph in rat. J. Lipid Res. **7**, 158–166 (1966).

BIER, D. M., HAVEL, R. J.: Activation of lipoprotein lipase by lipoprotein fractions of human serum. J. Lipid Res. **11**, 565–570 (1970).

BORGSTRÖM, B.: On the mechanism of the intestinal fat absorption. V. The effect of bile diversion on fat absorption in the rat. Acta Physiol. Scand. **28**, 279–286 (1953).

BORGSTRÖM, B.: The dimensions of the bile salt micelle. Measurements by gel filtration. Biochim. Biophys. Acta **106**, 171–183 (1965).

BRINDLEY, D. N., HÜBSCHER, G.: The effect of chain length on the activation and subsequent incorporation of fatty acids into glycerides by the small intestinal mucosa. Biochim. Biophys. Acta **125**, 92–105 (1966).

BROWN, V., LEVY, R. I., FREDRICKSON, D. S.: Studies of the proteins in human plasma very low density lipoproteins. J. Biol. Chem. **244**, 5687–5694 (1969).

BRUNZELL, J. D., HAZZARD, W. R., PORTE, D., JR., BIERMAN, E. L.: Evidence for a common, saturable, triglyceride removal mechanism for chylomicrons and very low density lipoproteins in man. J. Clin. Invest. **52**, 1578–1585 (1973).

CARDELL, R. R., JR., BADENHAUSEN, S., PORTER, K. P.: Intestinal triglyceride absorption in the rat. An electron microscopical study. J. Cell. Biol. **34**, 123–155 (1967).

CAREY, M. C., SMALL, D. M.: The characteristics of mixed micellar solutions with particular reference to bile. Amer. J. Med. **49**, 590–608 (1970).

CARTER, E. A., DRUMMEY, G. D., ISSELBACHER, K. J.: Ethanol stimulation of triglyceride synthesis by the intestine. Science **174**, 1245–1247 (1971).

CLARK, B., HÜBSCHER, G.: Biosynthesis of glycerides in subcellular fractions of intestinal mucosa. Biochim. Biophys. Acta **46**, 479–494 (1961).

CLARK, M. L., LANZ, H. C., SENIOR, J. R.: Bile salt regulation of fatty acid absorption and esterification in rat everted jejunal sacs *in vitro* and in thoracic duct lymph *in vivo*. J. Clin. Invest. **48**, 1587–1599 (1969).

DAINTY, J.: Water relations of plants. Adv. Botanical Res. **1**, 279–326 (1963).

DAWSON, A. M.: Bile salts and fat absorption. Gut **8**, 1–3 (1967).

DAWSON, A. M., ISSELBACHER, K. J.: Studies in lipid metabolism in the small intestine with observations on the role of bile salts. J. Clin. Invest. **39**, 730–740 (1960).

DEN BESTEN, L., REYNA, R. H., CONNOR, W. E., STEGINK, L. D.: The different effects on the serum lipids and fecal steroids of high carbohydrate diets given orally or intravenously. J. Clin. Invest. **52**, 1384–1393 (1973).

DIETSCHY, J. M.: Effects of bile salts on intermediate metabolism of the intestinal mucosa. Fed. Proc. **26**, 1589–1598 (1967).

DIETSCHY, J. M.: Mechanisms for the intestinal absorption of bile acids. J. Lipid Res. **9**, 297–309 (1968).

DOWLING, R. H.: Intestinal adaptation (editorial). New Eng. J. Med. **288**, 520–521 (1973).

FELDMAN, E. B., BORGSTRÖM, B.: Phase distribution of sterols: Studies by gel filtration. Biochim. Biophys. Acta **125**, 136–147 (1966).

FRANKS, J. J., RILEY, E. M., ISSELBACHER, K. J.: Synthesis of fatty acids by rat intestine *in vitro*. Proc. Soc. Exp. Biol. Med. **121**, 322–327 (1966).

GAGE, S. H., FISH, P. A.: Fat digestion, absorption, and assimilation in man and animals as determined by the dark-field microscope, and a fatsoluble dye. Amer. J. Anat. **34**, 1–85 (1924).

GALLAGHER, N., WEBB, J., DAWSON, A. M.: The absorption of ^{14}C-triolein in bile fistula rats. Clin. Sci. **29**, 73–82 (1965).

GANGL, A., OCKNER, R.: Intestinal metabolism of plasma free fatty acids. Clin. Res. **22** (3), 358 A (1974).

GLICKMAN, R. M., KIRSCH, K.: Lymph chylomicron formation during the inhibition of protein synthesis. J. Clin. Invest. **52**, 2910–2920 (1973).

GLICKMAN, R. M., KIRSCH, K., ISSELBACHER, K. J.: Fat absorption during inhibition of protein synthesis: Studies of lymph chylomicrons. J. Clin. Invest. **51**, 356–363 (1972).

GOODMAN, D. S.: The interaction of human erythrocytes with sodium palmitate. J. Clin. Invest. **37**, 1729—1735 (1958).

GOTTO, A. M., LEVY, R. I., JOHN, K., FREDRICKSON, D. S.: On the protein defect in abetalipoproteinemia. New Eng. J. Med. **284**, 813–818 (1971).

GREENBERGER, N. J., RODGERS, J. B., ISSELBACHER, K. J.: Absorption of medium and long chain triglycerides: Factors influencing their hydrolysis and transport. J. Clin. Invest. **45**, 217–227 (1966).

GRUNDY, S., AHRENS, E. H., SALEN, G.: Hourly rates of secretion of biliary lipids in man. Clin. Res. **18**, 529 (1970).

HAESSLER, H. A., ISSELBACHER, K. J.: The metabolism of glycerol by intestinal mucosa. Biochim. Biophys. Acta **73**, 427–436 (1963).

HARDISON, W. G. M., ROSENBERG, I. H.: Bile salt deficiency in the steatorrhea following resection of the ileum and colon. New Eng. J. Med. **277**, 337–342 (1967).

HAVEL, R. J.: Pathogenesis, differentiation and management of hypertriglyceridemia. Adv. Int. Med. **15**, 117–154 (1969).

HAVEL, R. J., KANE, J. P., KASHYOP, M. L.: Interchange of apolipoproteins between chylomicrons and high density lipoproteins during alimentary lipemia in man. J. Clin. Invest. **52**, 32–38 (1973).

HEIMBERG, M., WILCOX, H. G.: The effect of palmitic and oleic acids on the properties and composition of the very low density lipoprotein secreted by the liver. J. Biol. Chem. **247**, 875–880 (1972).

HOFMANN, A. F.: Micellar solubilization of fatty acids and monoglycerides by bile salt solutions. Nature **190**, 1106–1107 (1961).

HOFMANN, A. F., BORGSTRÖM, B.: Physico-chemical state of lipids in intestinal content during their digestion and absorption. Fed. Proc. **21**, 43–50 (1962).

HOFMANN, A. F., BORGSTRÖM, B.: The intralumenal phase of fat digestion in man: The lipid content of the micellar and oil phases of intestinal content during fat digestion and absorption. J. Clin. Invest. **43**, 247–257 (1964).

HOFMANN, A. F., KERN, F., JR.: The significance of bile acids in gastrointestinal and hepatic disease. In: Disease-a-Month, Year Book Medical Publishers, Inc., 1–38 (1971).

HOFMANN, A. F., MEKHJIAN, H. S.: Bile acids and the intestinal absorption of fat and electrolytes in health and disease. In: The bile acids (Vol. II). New York: Plenum Press 1973.

HOFMANN, A. F., POLEY, J. R.: Cholestyramine treatment of diarrhea associated with ileal resection. New Eng. J. Med. **281**, 397–402 (1969).

HOFMANN, A. F., SMALL, D. M.: Detergent properties of bile salts: Correlation with physiological function. Ann. Rev. Med. **18**, 333–376 (1967).

HOLT, P. R.: Intestinal absorption of bile salts in the rat. Amer. J. Physiol. **207**, 1–7 (1964).

HOVING, J., VALKEMA, A. J.: Effect of dietary fat content on the site of fat absorption in hamster small intestine. Biochim. Biophys. Acta **187**, 53–58 (1969).

HÜBSCHER, G., BRINDLEY, D. N., SMITH, M. E., SEDGWICK, B.: Stimulation of biosynthesis of glyceride. Nature **216**, 449–453 (1967).

HÜBSCHER, G., CLARK, B., WEBB, M. E., SHERRALT, H. S. A.: Structural and enzymatic relationship sin intestinal fat absorption. In: 7th International Conference on Biochemistry of Lipids: Biochemical problems of lipids, p. 201–210 (A. C. FRAZER, Ed.) Elsevier 1963.

ISENBERG, J. I., WALSH, J. H., GROSSMAN, M. I.: Zollinger-Ellison syndrome. Gastroenterology **65**, 140–165 (1973).

ISSELBACHER, K. J.: Biochemical aspects of lipid malabsorption. Fed. Proc. **26**, 1420–1425 (1967).

ISSELBACHER, K. J., SCHEIG, R., PLOTKIN, G. R., CAULFIELD, J. B.: Congenital β-lipoprotein deficiency: An hereditary disorder involving a defect in the absorption and transport of lipids. Medicine **43**, 347–361 (1964).

ITO, S.: The enteric coat on cat intestinal microvilli. J. Cell Biol. **27**, 475–491 (1965).

JERSILD, R. A., JR.: A time sequence study of fat absorption in the rat jejunum. Amer. J. Anat. **118**, 135–162 (1966).

JOHNSTON, J. M.: Mechanism of fat absorption. In: Handbook of physiology, vol. III, Sect. 6, 1353–1375, Amer. Physiol. Soc., Washington, D.C., 1968.

JOHNSTON, J. M., BORGSTRÖM, B.: The intestinal absorption and metabolism of micellar solutions of lipids. Biochim. Biophys. Acta **84**, 412–423 (1964).

JOHNSTON, J. M., PAULTAUF, F., SCHILLER, C. M., SCHULTZ, L. D.: The utilization of the α-glycerophosphate and monoglyceride pathways for phosphatidyl choline biosynthesis in the intestine. Biochim. Biophys. Acta **218**, 124–133 (1970).

JOHNSTON, J. M., RAO, G. A.: Triglyceride biosynthesis in the intestinal mucosa. Biochim. Biophys. Acta **106**, 1–9 (1965).

JOHNSTON, J. M., RAO, G. A., LOWE, P. A.: The separation of the α-glycerophosphate and monoglyceride pathways in the intestinal biosynthesis of triglycerides. Biochim. Biophys. Acta **137**, 578–580 (1967).

JOHNSTON, J. M., RAO, G. A., LOWE, P. A., SCHWARZ, B. E.: The nature of the stimulatory role of the supernatant fraction on triglyceride synthesis by the α-glycerophosphate pathway. Lipids **2**, 14–20 (1967).

JONES, A. L., OCKNER, R. K.: An electron microscopic study of endogenous very low density lipoprotein production in the intestine of rat and man. J. Lipid Res. **12**, 580–589 (1971).

JONES, A. L., RUDERMAN, N., HERRERA, G.: An electron microscopic study of lipoprotein production and release by the isolated perfused rat liver. Proc. Soc. Exp. Biol. Med. **123**, 4–9 (1966).

JONES, A. L., RUDERMAN, N., HERRERA, G.: An electron microscopic and biochemical study of lipoprotein synthesis in the isolated perfused rat liver. J. Lipid Res. **8**, 429–446 (1967).

KARMEN, A., WHYTE, M., GOODMAN, D. S.: Fatty acid esterification and chylomicron formation during fat absorption. I. Triglyceride and cholesterol esters. J. Lipid Res. **4**, 312–321 (1963).

KAYDEN, H. J., MEDICK, M.: The absorption and metabolism of short and long chain fatty acids in puromycin-treated rats. Biochim. Biophys. Acta **176**, 37–43 (1969).

KAYDEN, H. J., SENIOR, J. R., MATTSON, F. H.: The monoglyceride pathway of fat absorption in man. J. Clin. Invest. **46**, 1695–1703 (1967).

KORNBERG, A., PRICER, W. E., JR.: Enzymatic esterification of α-glycerophosphate by long chain fatty acids. J. Biol. Chem. **204**, 345–357 (1953).

KUHL, W. E., SPECTOR, A. A.: Uptake of long-chain fatty acid methyl esters by mammalian cells. J. Lipid Res. **11**, 458–465 (1970).

LAURENT, T. C., PERSSON, H.: A study of micelles of sodium taurodeoxycholate in the ultracentrifuge. Biochim. Biophys. Acta **106**, 616–624 (1965).

Levi, A. J., Gatmaitan, Z., Arias, I. M.: Two hepatic cytoplasmic protein fractions, Y and Z, and their possible role in the hepatic uptake of bilirubin, sulfobromophthalein, and other anions. J. Clin. Invest. **48**, 2156–2167 (1969).

Lo, C., Marsh, J. B.: Biosynthesis of plasma lipoproteins. Incorporation of ^{14}C-glucosamine by cells and subcellular fractions of rat liver. J. Biol. Chem. **245**, 5001–5006 (1970).

Mansbach, C. M. II: Complex lipid synthesis in hamster intestine. Biochim. Biophys. Acta **296**, 386–400 (1973).

Mishkin, S., Yalovsky, M., Kessler, J. I.: Stages of uptake and incorporation of micellar palmitic acid by hamster proximal intestinal mucosa. J. Lipid Res. **13**, 155–168 (1972).

Mistilis, S. P., Ockner, R. K.: Effects of ethanol on endogenous lipid and lipoprotein metabolism in small intestine. J. Lab. Clin. Med. **80**, 34–46 (1972).

Mukerjee, P.: Dimerization of anions of long-chain fatty acids in aqueous solutions and the hydrophobic properties of the acids. J. Phys. Chem. **69**, 2821–2827 (1965).

Ockner, R. K., Hughes, F. B., Isselbacher, K. J.: Very low density lipoproteins in intestinal lymph: origin, composition and role in lipid transport in the fasting state. J. Clin. Invest. **48**, 2079–2088 (1969a).

Ockner, R. K., Hughes, F. B., Isselbacher, K. J.: Very low density lipoproteins in intestinal lymph: role in triglyceride and cholesterol transport during fat absorption. J. Clin. Invest. **48**, 2367–2373 (1969b).

Ockner, R. K., Jones, A. L.: An electron microscopic and functional study of very low density lipoproteins in intestinal lymph. J. Lipid Res. **11**, 284–292 (1970).

Ockner, R. K., Manning, J.: Fatty acid binding protein in small intestine: identification, isolation and evidence for its role in cellular fatty acid transport. J. Clin. Invest. **54**, 326–338 (1974).

Ockner, R. K., Manning, J. M., Poppenhausen, R. B., Ho, W. K. L.: A binding protein for fatty acids in cytosol of intestinal mucosa, liver, myocardium, and other tissues. Science **177**, 56–58 (1972b).

Ockner, R. K., Pittman, J. P., Yager, J. L.: Differences in the intestinal absorption of saturated and unsaturated long chain fatty acids. Gastroenterology **62**, 981–992 (1972a).

Palay, S. L., Karlin, L. J.: An electron microscopic study of the intestinal villus. II. The pathway of fat absorption. J. Biophys. Biochem. Cytol. **5**, 373–384 (1959).

Polheim, D., David, J. S. K., Schultz, F. M., Wylie, M. B., Johnston, J. M.: Regulation of triglyceride biosynthesis in adipose and intestinal tissue. J. Lipid Res. **14**, 415–421 (1973).

Porter, H. P., Saunders, D. R.: Isolation of the aqueous phase of human intestinal contents during the digestion of a fatty meal. Gastroenterology **60**, 997–1007 (1971).

Porter, H. P., Saunders, D. R., Tytgat, G., Brunser, O., Rubin, C. E.: Fat absorption in bile fistula man. Gastroenterology **60**, 1008–1019 (1971).

Rao, G. A., Johnston, J. M.: Purification and properties of triglyceride synthetase from the intestinal mucosa. Biochim. Biophys. Acta **125**, 465–473 (1966).

Rao, G. A., Johnston, J. M.: Studies of the formation and utilization of bound CoA in glyceride biosynthesis. Biochim. Biophys. Acta **144**, 25–33 (1967).

Redgrave, T. G., Zilversmit, D. B.: Does puromycin block release of chylomicrons from the intestine? Amer. J. Physiol. **217**, 336–340 (1969).

Rodgers, J. B., Jr., Bochenek, W.: Localization of lipid re-esterifying enzymes of the rat small intestine. Effects of jejunal removal on ileal enzyme activities. Biochim. Biophys. Acta **202**, 426–435 (1970).

Roheim, P. S., Gidez, L. I., Eder, H. A.: Extrahepatic synthesis of lipoproteins of plasma and chyle: role of the intestine. J. Clin. Invest. **45**, 297–300 (1966).

Sabesin, S. M., Isselbacher, K. J.: Protein synthesis inhibition: mechanism for the production of impaired fat absorption. Science **147**, 1149–1151 (1965).

Salpeter, M. M., Zilversmit, D. B.: The surface coat of chylomicrons: electron microscopy. J. Lipid Res. **9**, 187–192 (1968).

Salt, H. B., Wolff, O. H., Lloyd, J. K., Fosbrooke, A. S., Cameron, A. H., Hubble, D. V.: On having no beta-lipoprotein. A syndrome comprising abeta-lipoproteinemia, acanthocytosis, and steatorrhea. Lancet **2**, 325–329 (1960).

Schafer, J. A., Andreoli, T. E.: Water transport in biological and artificial membranes. Arch. Int. Med. **129**, 279–292 (1972).

Scow, R. O.: Incorporation of dietary lecithin and lysolecithin into lymph chylomicrons in the rat. J. Biol. Chem. 242, 4919–4924 (1967).

Senior, J. R., Isselbacher, K. J.: Formation of higher glycerides from monopalmitin and palmityl-CoA by microsomes of rat intestinal mucosa. Biochem. Biophys. Res. Comm. 6, 274–278 (1961).

Senior, J. R., Isselbacher, K. J.: Direct esterification of monoglycerides with palmityl coenzyme A by intestinal epithelial subcellular fractions. J. Biol. Chem. 237, 1454–1459 (1962).

Senior, J. R.: Intestinal absorption of fats. J. Lipid Res. 5, 495–521 (1964).

Shimoda, S. S., Saunders, D. R., Rubin, C. E.: The Zollinger-Ellison syndrome with steatorrhea. II. The mechanisms of fat and vitamin B_{12} malabsorption. Gastroenterology 55, 705–723 (1968).

Shohet, S., Nathan, D. G., Karnovsky, M. L.: Stages in the incorporation of fatty acids into red blood cells. J. Clin. Invest. 47, 1096–1108 (1968).

Singh, A., Balint, J. A., Edmonds, R. H., Rodgers, J. B.: Adaptive changes of the rat small intestine in response to a high fat diet. Biochim. Biophys. Acta 260, 708–715 (1972).

Small, D. M.: A classification of biologic lipids based upon their interaction in aqueous systems. J. Amer. Oil Chem. Soc. 45, 108–119 (1968).

Small, D. M.: Surface and bulk interactions of lipids and water with a classification of biologically active lipids based on these interactions. Fed. Proc. 29, 1320–1326 (1970).

Small, D. M., Penkett, S. A., Chapman, D. M.: Studies on simple and mixed bile salt micelles by nuclear magnetic resonance spectroscopy. Biochim. Biophys. Acta 176, 178–189 (1969).

Smith, R., Tanford, C.: Hydrophobicity of long chain n-alkyl carboxylic acids, as measured by their distribution between heptane and aqueous solutions. Proc. Nat. Acad. Sci. 70, 289–293 (1973).

Spector, A. A., Steinberg, D., Tanaka, A.: Uptake of free fatty acids by Ehrlich ascites tumor cells. J. Biol. Chem. 240, 1032–1041 (1965).

Steim, J. M., Edner, O. J., Bargoot, F. G.: Structure of human serum lipoproteins: nuclear magnetic resonance supports a micellar model. Science 162, 909–911 (1968).

Strauss, E. W.: Morphological aspects of triglyceride absorption. In: Handbook of physiology, Vol. III, Sect. 6, pp. 1377–1406. Amer. Physiol. Soc., Washington, D.C. 1968.

Tilney, L. G., Mooseker, M.: Actin in the brush border of epithelial cells of the chick intestine. Proc. Nat. Acad. Sci. 68, 2611–2615 (1971).

Trier, J. S.: Morphology of the epithelium of the small intestine. In: Handbook of physiology, Vol. III, Sect. 6, pp. 1125–1175. Amer. Physiol. Soc., Washington, D.C. 1968.

Trier, J. S.: The surface coat of gastrointestinal epithelial cells. Gastroenterology 56, 618–622 (1969).

Tytgat, G. N., Rubin, C. E., Saunders, D. R.: Synthesis and transport of lipoprotein particles by intestinal absorptive cells in man. J. Clin. Invest. 50, 2065–2078 (1971).

Van Deest, B. W., Fordtran, J. S., Morawski, S. G., Wilson, J. D.: Bile salt and micellar fat concentration in proximal small bowel contents of ileectomy patients. J. Clin. Invest. 47, 1314–1324 (1968).

Ways, P. O., Parmentier, C. M., Kayden, H. J., Jones, J. W., Saunders, D. R., Rubin, C. E.: Studies on the absorptive defect for triglyceride in abetalipoproteinemia. J. Clin. Invest. 46, 35–46 (1967).

Westergaard, H., Dietschy, J. M.: Structure of the unstirred water layer in the intestine. Clin. Res. 20, 736 (1972).

Wilson, F. A., Dietschy, J. M.: Characterization of bile acid absorption across the unstirred water layer and brush border of the rat jejunum. J. Clin. Invest. 51, 3015–2035 (1972).

Wilson, F. A., Sallee, V. L., Dietschy, J. M.: Unstirred water layers in intestine: rate determinant of fatty acid absorption from micellar solutions. Science 174, 1031–1033 (1971).

Windmueller, H. G., Lindgren, F. T., Lossow, J., Levy, R. I.: On the nature of circulatory lipoproteins of intestinal origin in the rat. Biochim. Biophys. Acta 202, 507–516 (1970).

Windmueller, H. G., Spaeth, A. E.: Fat transport and lymph and plasma lipoprotein biosynthesis by isolated intestine. J. Lipid Res. 13, 92–105 (1972).

Zilversmit, D. B.: The surface coat of chylomicrons: lipid chemistry. J. Lipid Res. 9, 180–186 (1968).

Author Index

Page numbers in *italics* refer to the bibliography

Subject Index

Reviews of
Physiology,
Biochemistry
and Experimental
Pharmacology

Vol. 60: 48 figures
IV, 401 pages
(6 pages in German)
1968. Cloth DM 120,—
ISBN 3-540-04103-6

Contents: R. T. Kelleher
and W. H. Morse: Deter-
minants of the Specificity
of Behavioral Effects of
Drugs. — B. Jeanrenaud:
Adipose Tissue
Dynamics and
Regulation, Revisited. —
P. P. Foà: Glucagon. —
E. Habermann: Bio-
chemie, Pharmakologie
und Toxikologie der
Inhaltsstoffe von
Hymenopterengiften.

Vol. 61: 69 figures
1 portrait. III, 291 pages
(104 pages in German)
1969. Cloth DM 110,—
ISBN 3-540-04472-8

Contents: R. Jung:
Paul Hoffmann 1884-
1962. W. Ulbricht:
The Effect of Veratridine
on Excitable Membranes
of Nerve and Muscle. —
M. Wiesendanger: The
Pyramidal Tract. Recent
Investigations on its Mor-
phology and Function. —
M. E. Holman: Electro-
physiology of Vascular
Smooth Muscle. —
O.-J. Grüsser,
U. Grüsser-Cornehls:
Neurophysiologie des
Bewegungssehens.

Vol. 62: 14 figures
1 portrait. VIII, 159 pages
(23 pages in German)
1970. Cloth DM 58,—
ISBN 3-540-04811-1

Contents: M. Vogt:
John Henry Gaddum,
1900-1965. —
J. M. Marshall:
Adrenergic Innervation
of the Female
Reproductive Tract:
Anatomy, Physiology
and Pharmacology. —
W.-D. Thomitzek:
Die Rolle des Carnitins
im Intermediärstoff-
wechsel. — F. Bergel:
Today's Carcino-
chemotherapy: Some of
its Achievements,
Failures and Prospects.

Vol. 63: 79 figures
1 portrait
VIII, 241 pages. 1971
Cloth DM 110,—
ISBN 3-540-05317-4

Contents: H. O. Schild:
Henry Hallett Dale, 1875-
1968. — R. F. Schmidt:
Presynaptic Inhibition
in the Vertebrate
Central Nervous
System. — W. Schaper:
The Physiology of the
Collateral Circulation
in the Normal and
Hypoxic Myocardium. —
H. Schmid-Schönbein,
R. E. Wells, jr.:
Rheological Properties
of Human Erythrocytes
and their Influence
upon the "Anomalous"
Viscosity of Blood.

**Vol. 64: Neuro-
physiology and Neuro-
chemistry of sleep and
wakefulness**
Vergriffen

Vol. 65: 45 figures
1 portrait
VIII, 191 pages. 1972
Cloth DM 75,—
ISBN 3-540-05814-1

Contents: P. Kruhøffer,
Chr. Crone: Einar Lunds-
gaard, 1899-1968. —
J.-S. Pitton: Mechanisms
of Bacterial Resistance
to Antibiotics. —
J. B. Stanbury: Some
Recent Developments
in the Physiology of
the Thyroid Gland. —
M. A. Bouman,
J. J. Koenderink:
Psychophysical Basis of
Coincidence
Mechanisms in the
Human Visual System.

Prices are subject to
change without notice

Springer-Verlag
Berlin
Heidelberg
New York
München Johannesburg
London Madrid
New Delhi Paris
Rio de Janeiro Sydney
Tokyo Utrecht Wien

Reviews of Physiology, Biochemistry and Experimental Pharmacology

Vol. 66: 46 figures
1 portrait
VIII, 296 pages. 1972
Cloth DM 110,—
ISBN 3-540-05882-6

Contents:
H.-J. Schümann:
Peter Holtz, 1902-
1970. — H. van den Bosch,
L. M. G. van Golde,
L. L. M. van Deenen:
Dynamics of Phospho-
glycerides. —
N. Emmelin,
U. Trendelenburg:
Degeneration Activity
after Parasympathetic
or Sympathetic Dener-
vation. — P. N. Patil,
J. B. LaPidus: Stereo-
isomerism of Adrenergic
Drugs.

Vol. 67: 43 figures
IV, 226 pages. 1972
Cloth DM 96,—
ISBN 3-540-05959-8

Contents: K. Koizumi,
C. M. Brooks: The Inte-
gration of Autonomic
Reactions: A Discussion
of Autonomic Reflexes,
their Control and their
AssociationwithSomatic
Reactions. — W. Elger:
Physiology and Pharma-
cology of Female
Reproduction under the
Aspect of Fertility
Control. —
W. H. Daughaday,
L. S. Jacobs: Human
Prolactin.

Vol. 68: 12 figures
IV, 128 pages. 1973
Cloth DM 75,—
ISBN 3-540-06238-6

Contents: P. A. Owren,
H. Stormorken: The
Mechanism of Blood
Coagulation. —
W. W. Fleming,
J. J. McPhillips,
D. P. Westfall: Post-
junctional Supersensi-
tivity and Subsensitivity
of Excitable Tissues
to Drugs. — R. S. Turner,
M. M. Burger: The Cell
Surface in Cell Inter-
actions.

Vol. 69: 15 figures
IV, 222 pages. 1974
Cloth DM 98,—
ISBN 3-540-06498-2

Contents: J. B. Furness,
M. Costa: The Adren-
ergic Innervation of the
Gastrointestinal Tract.
— S. Numa: Regulation
of Fatty-Acid Synthesis
in Higher Animals. —
D. R. Curtis, G. A. R.
Johnston: Amino Acid
Transmitters in the
Mammalian Central
Nervous System.

Vol. 70: 23 figures
IV, 174 pages. 1974
Cloth DM 88,—
ISBN 3-540-06716-7

Contents: C. Bauer:
On the Respiratory
Function of Haemo-
globin. — M. P. Blau-
stein: The Interrela-
tionship between
Sodium and Calcium
Fluxes across Cell
Membranes. —
H. Blaschko: The
Natural History of
Amine Oxidases.

Prices are subject to
change without notice

Springer-Verlag
Berlin
Heidelberg
New York

München Johannesburg
London Madrid
New Delhi Paris
Rio de Janeiro Sydney
Tokyo Utrecht Wien